Wissenschaftliche Beiträge zur Medizinelektronik

Band 2

Wissenschaftliche Beiträge zur Medizinelektronik

Band 2

Herausgegeben von

Wolfgang Krautschneider

Low-Noise and Low-Power CMOS Amplifiers, Analog Front-Ends, and Systems for Biomedical Signal Acquisition

Vom Promotionsausschuss der
Technischen Universität Hamburg-Harburg

zur Erlangung des akademischen Grades

Doktor-Ingenieur (Dr.-Ing.)

genehmigte Dissertation

von
Jakob Martin Tomasik

aus
Gliwice

2010

1. Gutachter: Prof. Dr.-Ing. Wolfgang Krautschneider
2. Gutachter: Prof. Dr.-Ing. Manfred Kasper

Tag der mündlichen Prüfung: 12. März 2010

Jakob M. Tomasik

Low-Noise and Low-Power CMOS Amplifiers, Analog Front-Ends, and Systems for Biomedical Signal Acquisition

Logos Verlag Berlin

λογος

Wissenschaftliche Beiträge zur Medizinelektronik

Herausgegeben von
Prof. Dr. Wolfgang Krautschneider

Technische Universität Hamburg-Harburg
Institut für Nanoelektronik
Eißendorfer Str. 38
D-21073 Hamburg

Bibliografische Information Der Deutschen Bibliothek

Die Deutsche Bibliothek verzeichnet diese Publikation in der Deutschen Nationalbibliografie; detaillierte bibliografische Daten sind im Internet über http://dnb.ddb.de abrufbar.

ISBN 978-3-8325-2481-4
ISSN 2190-3905

Logos Verlag Berlin GmbH
Comeniushof, Gubener Str. 47,
10243 Berlin

Tel.: +49 (0)30 / 42 85 10 90
Fax: +49 (0)30 / 42 85 10 92
http://www.logos-verlag.de

To
John, Angelika, Christine, Peter,
and Elisabeth

An opamp is a wild tiger that
needs to be put in a cage. The
feedback network is that cage.

Johan H. Huijsing

Danksagung

Zunächst möchte ich mich bei Prof. Dr. Wolfgang Krautschneider, Leiter des Instituts für Nanoelektronik an der Technischen Universität Hamburg-Harburg, für die hervorragende Betreuung und das mir entgegengebrachte Vertrauen sehr herzlich bedanken.

Ein ganz besonderer Dank gilt Wjatscheslaw (Slawa) Galjan für die großartige Teamarbeit und stetige Unterstützung im M3C-Projekt. Dies gilt auch für Kristian Hafkemeyer, der das Team am Ende des Projektes exzellent unterstützte. Allen beiden verdanke ich wertvolle Diskussionen und Anregungen, die zum Gelingen dieser Arbeit beigetragen haben.

I would also like to express my sincere appreciation to Mr. Nick van Helleputte for the great cooperation with the KU Leuven and the time spent at the various M3C meetings.

Für die fachliche Unterstützung und zahlreichen Diskussionen im Bereich der programmierbaren Operationsverstärker danke ich Herrn Dr.-Ing. Dietmar Schröder.

Herrn Prof. Dr. Manfred Kasper danke ich für die Zweitbegutachtung sowie Herrn Prof. Dr. Ernst Brinkmeyer für die Übernahme des Prüfungsvorsitzes. Weiterhin sei Frau Gabriele Heinrich für die sprachliche Unterstützung gedankt.

Meiner Familie, insbesondere meiner Frau Angelika, meinen Eltern und meiner Großmutter danke ich für die Geduld und Unterstützung während der Entstehung dieser Arbeit.

Jakob M. Tomasik, April 2010

Contents

Chapter 1

Introduction

In recent years, the research and development of CMOS integrated circuits for biomedical signal acquisition systems has increased significantly. This has resulted and will result in novel medical device solutions which target at both, new therapeutic treatments and lower health costs. Concurrently, the development of CMOS integrated circuits has reached the deep-submicron range with device sizes becoming a few nanometers in the near future.

The increase in research activity in the area of biomedical engineering, a superset of biomedical signal acquisition systems, can be impressively demonstrated by performing a web search for scientific literature which includes the word *Biomedical Engineering* and sorting the results for the year of articles published [1]. The result is illustrated in Fig. 1.1 for the year 1991 to 2007 and shows that the number of articles published increased by more than 1200 % within that period of time.

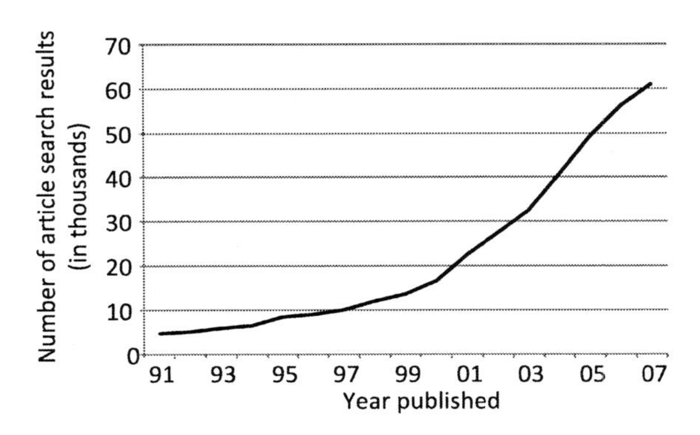

Figure 1.1: Number of *Google Scholar* search results for articles published in a year and containing the key word *Biomedical Engineering* [1].

What are the reasons for such a high increase in research activity within this area? One factor is the population aging in developed countries giving rise to increased health care and related costs. For example, the ratio of 60-year-old to newborn was approximately one in 2005 in Germany, whereas it is predicted that in 2050 this ratio will increase to be two [2]. Therefore, it is essential to reduce health costs, e.g. using a home care approach, and to introduce novel therapeutic approaches. On the other hand, the development of integrated circuits and packaging technologies led to a miniaturization of electronic devices which can be used to design portable, long-term operating medical devices or to use novel implant technologies. The level of miniaturization of digital CMOS integrated circuit technology at the time of writing has reached the 22 nm process technology with 2.9 billion transistors packed in a test chip [3].

This work is the result of a European Union research project to develop a multi-monitoring integrated circuit for homecare applications to provide a novel solution with respect to the aforementioned problems and challenges. The work describes the analysis, the design and measurement results of CMOS amplifiers, analog front-end and integrated system solutions for the acquisition of biomedical signals like ECG, EEG, EMG and EP. Additionally, the transition to deep-submicron analog circuits is presented by means of an amplifier design suited for applications within the framework of biomedical signal acquisition systems.

This work is organized as follows: Chapter 2 and 3 introduce the basics of biomedical signal acquisition and analog CMOS fundamentals, respectively. The analysis and implementation of low-power and low-noise amplifiers and analog front-end building blocks for use in biomedical signal acquisition systems is presented in Chapter 4. Using these results, the realization of two integrated system solutions including several signal acquisition channels is described in Chapter 5. Finally, conclusions and discussions are given in Chapter 6.

Chapter 2

Biomedical Signal Acquisition

The design of CMOS circuits for the acquisition of biomedical signals requires a knowledge of the medical background of the signals and sensing procedures. The medical background is needed to understand the origin of biopotentials and their characteristics at the location of measurement. The electrodes needed to sense the biomedical signals give additional constraints to the electronic system due to their physical properties and placement topologies required by the physicians.

This chapter gives an introduction to the origins and types of biomedical signals and to electrodes in Sections 2.1, 2.2 and 2.3, respectively. Section 2.4 gives an overview of constraints that have to be considered for the design of biomedical amplifiers. Finally, Section 2.5 describes the building blocks of a generic analog front-end (AFE) design.

2.1 The Origin of Bioelectricity

The bioelectric phenomena in the human body are caused by *excitable cells* like nerve or muscle cells. At steady state, these cells exhibit a *resting potential* between the inside of the cell and the extracellular fluids, both separated by a cell membrane that has different permeability to specific ions. The resting potential results from concentration imbalances of ions [4, 5], in particular sodium (Na^+), potassium (K^+) and chloride (Cl^-).

A mechanism called *sodium-potassium pump* which actively moves Na^+ ions out and K^+ ions into the cell at a rate of 3:2 creates a diffusion gradient for K^+ and Na^+. At rest, the permeability of the membrane to K^+ is approx. 100 times higher than its permeability to Na^+. Therefore, predominantly K^+ diffuses from the cell interior to the outside and an electric field builds up that tends to inhibit the outward flux of K^+. A relatively high extracellular Cl^- ion concentration supports the inflow of Cl^- ions, but this is limited by the electric field due to the K^+ concentration. At steady state, a balance of the electric and diffusive forces is reached. As a result, a negative voltage with respect to the extracellular fluids exists across the membrane, the cell is said to be *polarized.* The typical value for the resting potential is -70 mV.

Excitable cells respond to the stimulation by an electric field with an increase of permeability to Na^+ ions [6, 5]. These move inside the cell thereby *depolarizing* the cell potential which becomes positive. Shortly afterwards, this inflow is stopped by a reduced Na^+ permeability. Now, the permeability to K^+ ions increases and these ions move out of the cell *repolarizing* the cell potential. At the end of this process the cell potentials may reach values more negative than the resting potential which is called *hyperpolarization.* Hereafter, both permeabilities return to their equilibrium values, first for the Na^+ ions and subsequent for the K^+ ions. The resting state is restored by means of the sodium-potassium pump.

The resulting waveform or impulse is called *action potential* and has a positive voltage peak of approx. $20 - 40$ mV with respect to the extracellular media. This impulse can travel down the cell membrane and trigger adjacent excitable cells which leads to a signal propagation for the nervous system and muscle cells, e.g. heart muscle cells. A second source of electrical activity is the connection of neurons, a major class of excitable cells, to other neurons or cells. This connection is established at specialized junctions at the neuron to connect to other cells. These junctions are called *synapses.* If one side of this junction is triggered by an action potential, chemical messenger molecules called *neurotransmitter* are set free and move to the other side. Here they trigger a change in permeability at the opposite side of the junction. The potentials resulting from this are called *postsynaptic potentials (PSP).* The duration of the resulting waveform is much longer compared to action potentials. They normally last $10 - 50$ ms and have amplitudes in the range of 10 mV.

Using a simplified model, bioelectric signal sources can be considered as a *current dipole* in a volume conductor representing the surrounding tissue [7]. In order to describe a current dipole, first a current monopole is described. A *current monopole* can be thought of as being an idealized point source of total current I_0 within the volume conductor. A physical *current dipole* consists of two current monopoles separated by a small distance d and applying the same total current with opposite sign to the conductive volume. A mathematical dipole is defined for $d \to 0$, $I_0 \to \infty$ and keeping the product $I_0\, d$ constant. The level of modeling complexity determines if the current dipole represents a single ion channel, a nerve fiber or a cluster of excitable cells.

The current dipole creates a current density distribution within the conducting media which results in potential differences that are measured at the surface of the body using electrodes; Fig. 2.1 illustrates this concept in a simplified version. The actual impedance of the human body as a volume conductor and the signal sources are much more complex than this simple model. Therefore, the use of finite element modeling (FEM) of the body is needed in order to accurately obtain the voltage difference measured between the electrodes that are placed at the body surface [8].

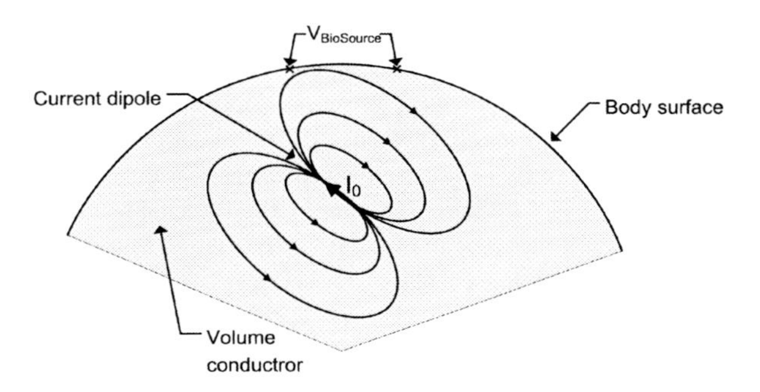

Figure 2.1: Simplified representation of a volume conductor field in the human body.

2.2 Biomedical Signals

The different types of bioelectric sources in the human body lead to a diversity in the form of biomedical signal acquisition. The main signal sources and associated types of signal acquisition are:

- Heart - Electrocardiogram (ECG)
- Brain - Electroencephalogram (EEG) / Evoked Potentials (EP)
- Muscle - Electromyogram (EMG)
- Nerve - Electroneurogram (ENG)
- Eye - Electroretinogram (ERG) /Electro-Oculogram (EOG)

The electrical activity of the human heart is recorded using an Electrocardiogram (ECG). A typical ECG waveform with specific features marked by letters is shown in Fig. 2.2. The heart is a muscle organ located in the thorax that pumps blood throughout the body [5, 9]. It is divided into a right and left side separated by the *septum*. Both sides of the heart contain two chambers, the *atrium* and the *ventricle*, each having valves to keep blood flowing just in one direction. The heartbeat is controlled by electrical active tissues, the *sinoatrial (SA) node* located at the right atrium, the *atrioventricular (AV) node* at the bottom of the septum of the atria and the *His-Purkinje system* located at the ventricle walls. The SA node serves as a pacemaker to the heart. During pumping, first the right atrium is filled with deoxygenated blood from the body and the left atrium is filled with oxygenated blood from the lung. When the right atrium is filled with blood the SA node located at the right atrium generates an electrical signal, which spreads across both atria and causes their contraction (P wave). This moves the blood into the corresponding ventricle. When the electric excitation reaches the AV node it is shortly delayed and then moves to the bundle of HIS and separates to the right and left bundle branches (Q wave). It then reaches the Purkinje fibers in the right and left ventricle walls which contract. The right ventricle pumps the deoxygenated blood to the lung and the left one pumps the oxygenated blood to the body. Again, valves at ventricle exit keep the blood from flowing back. The left ventricle contracts shortly (R wave) before the right ventricle (S wave). The pumping cycle ends with the relaxation of the ventricles (T wave).

The ECG signal is used for diagnosis of cardiovascular diseases like heart attack, arrhythmia and heart failure [9]. In order to obtain a spatial resolution, measurements from different combinations of electrode pairs are used, each of these pair combinations is called a *lead*. The standard 12-lead ECG measurement is based on three different lead configurations. The first configuration (3 leads), called *bipolar limb leads* uses electrodes connected to the right arm (RA), left arm (LA), right leg (RL) and left leg (LL). The right leg connection is used to ground the patient. Each lead of this configuration is obtained by measuring the difference between two electrodes, i.e. LA vs. RA (Lead I), LL vs. RA (Lead II) and LL vs. LA (Lead III). For the second configuration (3 leads), called *unipolar or augment limb leads*, a lead is formed by connecting two of the above electrodes together using a resistive network and measurement with the third one with respect to this connected pair, e.g. RA vs. LA+LL. The third configuration (6 leads) uses six unipolar chest electrodes and the potential are measured with respect to the three limb electrodes summed together by a resistive network. There also exists a configuration using less electrodes, e.g. for mobile ECG application.

The electrical activity of the human brain is recorded using the Electroencephalogram (EEG) [10, 11]. It can be recorded either on the scalp or directly from the surface of the brain. In addition, mircoelectrodes advanced into the brain can be used to improve the spatial resolution. The brain and the spinal cord which is connected to it form the *central nervous system.* The anatomy of the brain can be divided into different structures. The *brainstem* which is directly connected to the spinal cord is responsible for basic functions like control of heartbeat or respiration. The topmost part of the brainstem is the *diencephalon* which contains the

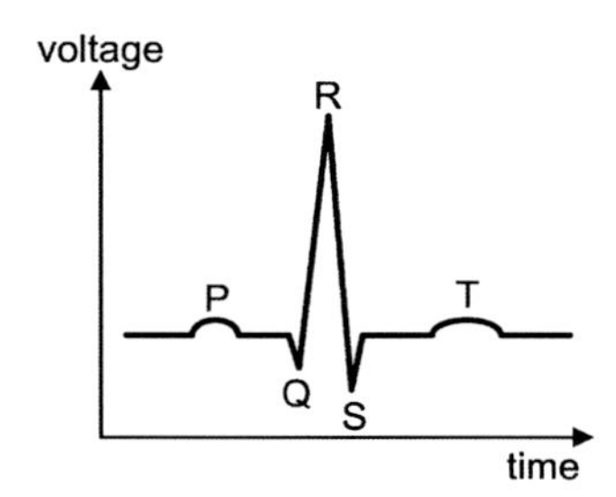

Figure 2.2: Sketch of a typical ECG waveform.

thalamus. The thalamus connects the sensory system to the enclosing *cerebrum*, the largest structure of the brain accounting for most of its surface area. The cerebrum and its outer layer, the *cerebral cortex* or just *cortex*, is responsible for higher brain functions like vision, hearing or speech. Located on the lower backside of the brain, the *cerebellum* controls the human muscle movement. The electrical activity measured on the scalp or surface of the brain originates almost entirely from the summed synchronous activity of many postsynaptic potentials in the cortex. Action potentials do not contribute to the measured EEG signals because of their relative short duration. The primary source of summed synchronous activity are the PSPs of *pyramidal cells.* This type of nerve cell has an elongated form oriented perpendicular to the surface of the brain. The parallel alignment of pyramidal cells in the cortex forms a basis for the summation of the electric fields. The rhythmic activity of these cells, mainly in times of a reduced stimulus from the associated sensory system, is believed to be controlled by the thalamus and structures related to it. The recorded EEG waveforms can be classified into four types as shown in Table 2.1 [11].

EEG Rhythm	Signal Bandwidth	Occurrence
Alpha (α)	8 Hz - 13 Hz	Relaxed state - Eyes closed
Beta (β)	13 Hz - 30 Hz	Mental activity
Gamma (γ)	$>$ 30 Hz	Information processing / Intentional motion
Delta (δ)	0.5 Hz - 4 Hz	Deep sleep
Theta (θ)	$>$ 4 Hz - 8 Hz	Emotional state

Table 2.1: Characteristics of EEG signals

A main application of the EEG is the examination and diagnosis of epileptic disorders. It is also often used in sleep diagnosis and recently for the use in brain-computer interfaces (BCI). The standard scheme to measure a scalp EEG is the

10-20 system. Four specific points on the human head are used as reference positions, the *nasion* (the transition point between nose and forehead), the *inion* (a characteristic swell at the occipital bone) and both ear lobes. Taking these points as references, electrodes are placed at a relative distance of 10% if one of the reference point is involved or 20% for all other electrode to electrode distances. This configuration uses 19 electrodes (21 including the references). For more improved source location *Dense Array EEG* with 256 or more electrodes can be used. The EEG signal traces are taken either bipolar, i.e. the voltage difference between two electrodes, or from one electrode with reference to an indifferent point, typically the ear lobe electrodes. In addition, patient grounding is needed. This is achieved by placing a ground electrode on the body, normally the first electrode above the nasion [10].

The interpretation of the EEG signals is rather difficult and requires a trained person. Nevertheless, one simple EEG event can be obtained easily by taking the EEG from an individual opening and closing his eyes. Here the signal taken from the backside of the head (*occipital region*) shows an increased α-wave activity by closing the eyes and a disappearance of α-wave activity by opening the eyes. This effect has been discovered by Hans Berger and has been accordingly named *Berger effect*. A α-wave recorded at the occipital region is shown in Fig. 2.3 [12].

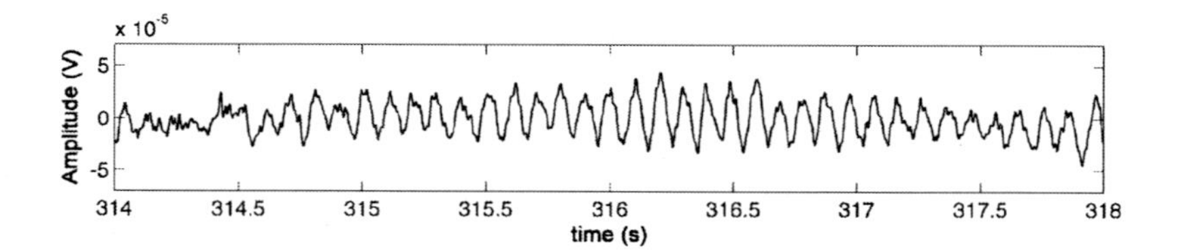

Figure 2.3: α-wave recorded at the occipital region in [12].

The recording of a change in EEG signal as a response to a controlled stimulus is called Evoked Potentials (EP) [5]. The EP is used for the diagnosis of dysfunctions along the pathways of the sensory system. The signal change in EEG is relative small in comparison to the background EEG activity, however the response will have a similar waveform every time the stimulation is repeated. This provides an opportunity to extract the provoked signal from the rather random background EEG by signal averaging. If we align each response of the measurement repeated n times synchronous with respect to the stimulation times, the signal-to-noise ratio (SNR) improves by a factor of $n/\sqrt{n} = \sqrt{n}$. The EP can be grouped into subclasses depending on the location of stimulus. For the Visual Evoked Potentials (VEP) the patient stares at a checkerboard pattern that periodically changes to its inverse. The VEP to this stimulus is recorded using an electrode placed at the occipital region (with respect to an reference electrode) and repeating the stimulus 100 times. In Brainstem Auditory Evoked Potentials (BAEP) a clicking sound is applied to one ear and the response is measured at the vertex with respect to a reference electrode. For Somatosensory Evoked Potentials (SSEP), a peripheral nerve is stimulated and the response is measured at the spinal cord and the scalp. This stimulus is repeated up to 1000 times.

The electric activity of the human skeletal muscles is recorded using the Electromyogram (EMG) [5]. The skeletal muscles consists of elongated fibers each containing substructure called *myofibrils* in a high number. The myofibrils are composed of two types of alternating filaments, *myosin filaments* and *actin filaments.* Muscle contraction is achieved by a sliding movement of the two filaments types in opposite direction. The contraction is initiated by an action potential within motor nerves that are connected to the muscle fiber via a synapse. If the action potential reaches the synapse, the neurotransmitter released induces an action potential in the muscle fiber membrane which travels along the fiber. This action potential triggers the sliding movement of the filaments. The EMG is used for diagnosis of nerve or muscle dysfunctions or as a control signal in prosthetics. It can be obtained by using surface electrodes or by needle electrodes inserted into the muscle. Needle electrodes have the advantage of a more localized signal pickup but bare a minimal risk for infection and discomfort for the patient.

The Electroneurogram (ENG) has similarities to the EMG, here nerve field potentials at two points along a peripheral nerve are measured to obtain the velocity of the signal [13]. The Electroretinogram (ERG) measures the electric response of the retina of the eye stimulated by light and the Electro-Oculogram measures the potential difference between cornea and retina to measure the direction of gaze [5]. Circuits and systems for the last three types of bioelectric measurements are not covered in this work, they have been listed here for completeness.

The biomedical signal types differ in their signal form at the location of measurement. The most important signal specifications used in signal processing are signal bandwidth and amplitude. Table 2.2 shows the associated signal characteristics based on [13, 14]

2.3 Electrodes

An important point in the characterization and design of biomedical signal acquisition systems is the interface between the human tissue and the electronic system.

Signal Type	Signal Bandwidth	Signal Amplitude
ECG	0.05 Hz - 250 Hz	5 μVpp - 8 mVpp
EEG	0.5 Hz - 100 (70[1]) Hz	2 μVpp - 200 μVpp
EMG	0.01 Hz - 5 kHz	50 μVpp - 10 mVpp
EP	0.1 Hz - 3 kHz	20 nVpp - 20 μVpp

[1] The IFCN standard for digital recording of clinical EEG [15] sets the cut-off frequency of the anti-aliasing filter to 70 Hz.

Table 2.2: Characteristics of biomedical signals

In order to translate the ionic currents into electric currents a transducer, in this case some form of electrode, is needed.

The simplest electrode for sensing biopotentials is composed of a metal in direct contact with the tissue which can be viewed as a solution containing ions, i.e. an electrolyte. Assuming an electrolyte with cations of the electrode material, an reduction-oxidation reaction at the metal/electrolyte interface occurs:

$$\mathrm{C_{ion}} \rightleftharpoons \mathrm{C_{ion}^{n+}} + \mathrm{nq} \tag{2.1}$$

$$\mathrm{A_{ion}^{m-}} \rightleftharpoons \mathrm{A_{ion}} + \mathrm{mq} \tag{2.2}$$

where q is the electron charge, and n and m are the cation (C_{ion}) and anion (A_{ion}) valences, respectively. Equations (2.1) and (2.2) describe how the conversion between ionic and electronic current is established between the metal and the electrolyte.

When a metal electrode is placed into an electrolyte containing the corresponding metal cations, reaction (2.1) starts to take place. Depending on the concentration of the cations in the electrolyte, either the reduction or the oxidation dominates. As a result, the electrolyte cation concentration at the metal/electrolyte interface changes which gives rise to a potential difference named *half-cell potential* [13]. The half-cell potential depends on the electrode material, the most popular electrode type uses silver chloride (AgCl) that has a rather low half-cell potential of +223 mV at room temperature.

The voltage across the metal-electrolyte interface may differ from the equilibrium half-cell potential by application of a current. This effect is mainly caused by a change of charge distribution at the electric double layer. The degree to which this effect alters the net charge transfer across the metal-electrolyte interface allows a differentiation between two electrode types [14]. For *polarizable electrodes* this effect is highly pronounced, hence displacement current dominates and the interface behaves more like a capacitor. Electrodes made from noble metal like platinum or gold, which have a low chemical reactivity, are an example of a polarizable electrode.

A very common type of *non-polarizable electrodes* is the silver-silver chloride

(Ag/AgCl) electrode. It is made up of three layers, a silver core or surface connected to the external wire, a thin silver chloride layer that is only slightly soluble in water and a surrounding electrolyte containing chloride anions. The silver discharges cations to the silver chloride layer leaving free electrons in the silver electrode and the electrolyte provides chloride anions that combine with the silver cations to extend the silver chloride layer. This type of non-polarizable electrode has a highly stable half-cell potential [13].

The Ag/AgCl electrodes have the disadvantage of skin irritations, drying-out of the electrolyte and low wearing comfort for long-term monitoring, i.e. starting from 24 hours up to years. Therefore, a different approach for sensing biopotentials has been taken recently by using capacitive electrodes (sometimes referred to as non-contact or insulated) [16, 17]. Here the electrode is separated by an insulating material, the biosignal is coupled capacitively from the tissue to the input terminal of the amplifier. On the one hand these types of electrodes improve the wearing comfort significantly, on the other hand they are more pronounced to motion artifacts, injected noise and demand a very high input impedance amplifier. Nevertheless, they seem to be the only suitable electrode type for long-time monitoring to this day.

Three representative designs of electrode types are shown in Fig. 2.4. The disposable, self-adhesive Ag/AgCl electrode is the most popular type for sensing of standard biopotentials at the skin surface and has been used for biomedical measurements described in this work (with an electrode diameter of about 1 cm). The micro-array needle-type electrode is used for implantable neuro-recordings. An example for this type of electrode is the Utah Electrode Array [18], it has an electrode count of 100 and measures $4 \times 4\text{mm}^2$. A exemplary setup of an capacitive electrode is shown in Fig. 2.4(c).

2.3.1 Electrical Characterization

In order to model the electrode-electrolyte interface electrically, an equivalent circuit is needed. The equivalent circuit should account for the given effects using passive components and voltage/current sources, nevertheless it can only be a rough estimation of the electric behavior due to the large extent of variability of the interface. A simple equivalent circuit of the electrode-electrolyte interface is given in Fig. 2.5 [19]. The DC (direct current) voltage source V_{ge} connected in series represents the half-cell potential. The resistance and capacitance of the metal (electrode) and the electrolyte (gel) interface are modeled by R_{ge} and C_{ge}, respectively. In addition, R_g has been included to represent the electrolyte resistance. This resistance does not apply for capacitive electrodes and is set to zero in such cases.

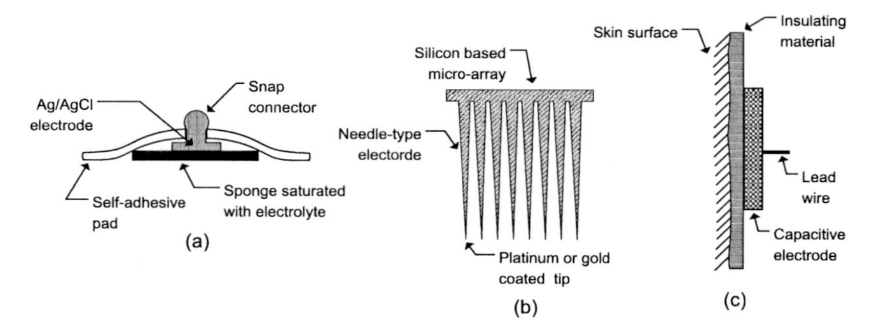

Figure 2.4: Different types of electrodes. (a) Disposable, self-adhesive Ag/AgCl electrode. (b) Silicon based micro-array of needle-type electrodes. (c) Capacitive (non-contact) electrode.

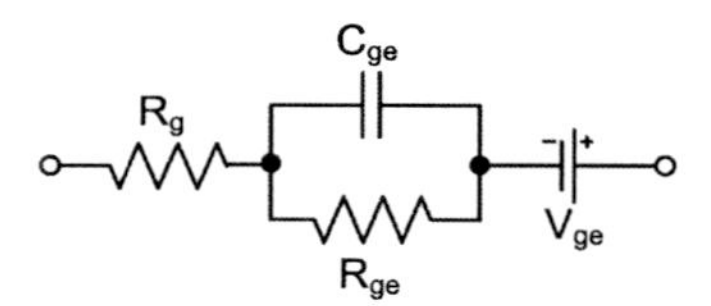

Figure 2.5: Basic electrode/electrolyte interface equivalent circuit.

The magnitude of the impedance of the equivalent circuit is approximately $R_g + R_{ge}$ at low frequencies, then starts to decrease as C_{ge} becomes effective and levels off at a value of R_g for high frequencies. The values for the circuit components in Fig. 2.5 vary substantially with respect to electrode and electrolyte materials used, the size and form of the electrode and voltage or current applied across the interface. The standardization for pregelled, disposable ECG electrodes by the Association for Advancement of Medical Instrumentation (AAMI) [20] defines a maximum impedance of 2 kΩ at 10 Hz for pregelled, disposable ECG electrode pairs that are connected gel-to-gel[1].

To fully characterize the behavior of the electrode placed onto the patient's body, an additional equivalent circuit is needed for the skin. The electrical properties of the skin resemble those of the metal-electrolyte interface. A potential difference V_{sg}, a capacitance C_{sg} and a resistance R_{sg} are associated with the skin-electrolyte

[1] The exact AAMI regulation states, that the average impedance value of at least 12 gel-to-gel connected electrodes should not exceed 2 kΩ at 10 Hz and none of the individual pair impedance should exceed 3 kΩ.

interface, respectively [13]. The potential difference and the capacitance arise mainly from the outermost skin layer, the *stratum corneum*, which consists of dead cells that originate from underlying layers. The resistance of the skin which is caused by sweat glands and other pathways and is added in parallel to the capacitance. Typical values for the skin-electrolyte interface are +10 to -70 mV for the potential difference V_{sg} [21] and $0.02 - 0.06\ \mu F \cdot cm^{-2}$ for the capacitance C_{sg} [22]. The skin resistance depends strongly on patient and skin preparation, it can have a value of $10\ \text{k}\Omega \cdot cm^{-2}$ to $5\ \text{M}\Omega \cdot cm^{-2}$ using a gelled ECG electrode (10 minutes after application) [22]. An extended electrode equivalent circuit including also the biomedical signal source, its internal impedance (see also Fig. 2.1) and a differential amplifier with input resistance R_{ain} and input capacitance C_{ain} is shown in Fig. 2.6 [23].

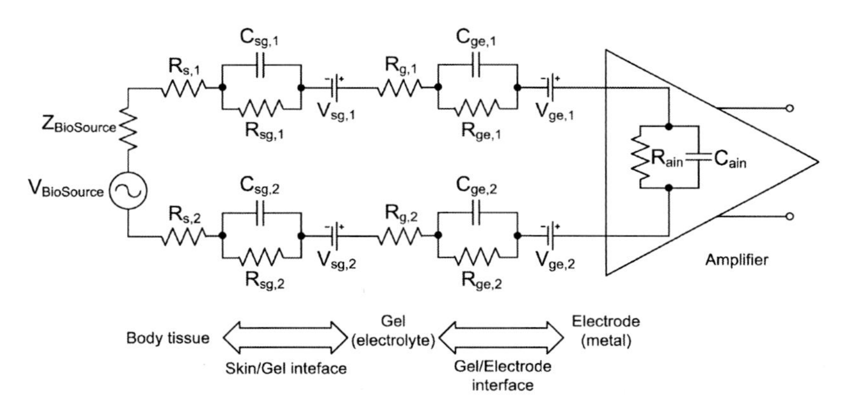

Figure 2.6: Extended electrode equivalent circuit.

The use of a differential amplifier for electrodes attached to the skin surface (i.e. standard Ag/AgCl electrodes) should theoretically cancel out the half-cell potentials. The ratio of the electrode half-cell voltages to those of the biomedical signals can easily reach values of more than 10,000 for small biomedical signals. Hence, any mismatch between the electrode pairs leads to a DC offset voltage $V_{DC,el}$ that is present at the amplifiers input. This applies also to the potential difference of the skin making the effect furthermore pronounced. The change of the electrode-gel and gel-skin interface potentials can be induced by large time constant effects like drying out of the gel or sweating by the patient. This results in a slow change of the DC voltage superposed on the biopotential signal and is referred to as baseline drift. An often encountered source of short time constant potential difference changes are motion artifacts. A reduction of these artifacts can be obtained by using an abrasive gel to remove the stratum corneum [24], a procedure that is often used in EEG measurements. The use of an abrasive gel has also the advantageous effect of lowering the skins resistance. Nevertheless, its use is quite unpleasant for the patient why any alternative to this seems quite attractive. Short time constant changes of $V_{DC,el}$ can also be induced by treatments like electrical stimulation or defibrillation of the patient.

Each of the above cases of electrode DC offset voltage $V_{DC,el}$ can lead to a saturation of the amplifier considering the amplification factor. Therefore, the DC offset voltage (and the difference of this voltage between two electrodes) has to be as small as possible. The AAMI standard for disposable ECG electrodes [20] requires a maximum DC offset voltage of ± 100 mV after 1 minute stabilization for a gel-to-gel connected pair of electrodes. A stimulation or electrode replacement exceeds this value temporarily until the signal is restored again. The time it takes for the signal to settle down depends not only on the time constant of the electrode but also on the time constant of the amplifier system, e.g. if a high pass filter is used. In any case, a temporary loss of the biomedical signal might be a severe disadvantage if physiological relevant events occur at these times. This necessitates a solution at the amplifier system or the use of capacitively coupled electrodes.

2.4 Integrated Analog Front-Ends

The main task of modern, integrated biomedical signal acquisition systems is to provide an accurate digital representation of biopotentials present at the electrodes. This representation is used for further processing or shown directly on a display. Prior to analog-to-digital conversion, amplification of the weak bioelectric signals is needed to exploit the full analog-to-digital converter (ADC) resolution. The amplifier acts also as an impedance transformer to prevent any loading of the bioelectric signal source. Usually, a difference amplifier is employed in a first front-end gain stage to amplify the difference between two electrode signals. In addition to this basic signal chain, additional signal conditioning might become necessary as a result of problems frequently encountered such as noise, interference or distortion. A simplified bioelectric signal chain, including interferences and non-idealities, is shown in Fig. 2.7 [13].

The signal chain non-idealities can be divided into three locations of origin

- Signal transduction and input leads
- Amplification

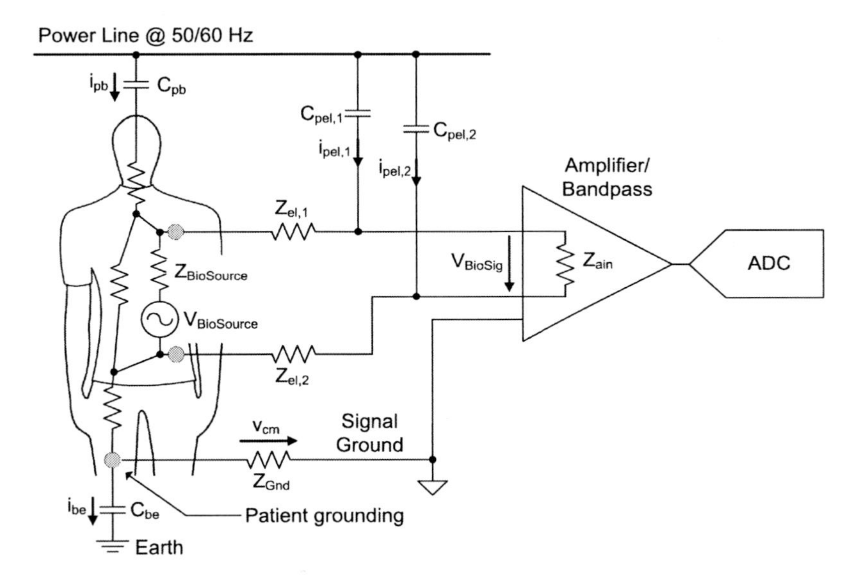

Figure 2.7: Biomedical signal acquisition (after [13]). The unlabeled impedances indicate the (distributed) body impedance.

- Analog-to-digital conversion

The biopotential signal V_{BioSig} at the input of the amplifier differs from the biopotential signal on the body $V_{BioSource}$ due to its internal impedance $Z_{BioSource}$, the electrode impedances $Z_{el,1/2}$ and a finite amplifier input impedance Z_{ain}. The specified biomedical signal peak-to-peak voltages at the amplifier input (V_{BioSig}), as given in Table 2.2, account for the effects of using application specific electrodes with the exemption of electrode DC-offsets.

Assuming no interferences, the overall gain of the biomedical signal acquisition systems, A, is defined by the maximum biopotential signal at the input, thus

$$A = \frac{V_{DD}}{\max\{V_{BioSig,pp}\}} \tag{2.3}$$

with V_{DD} being the amplifier and ADC supply voltage. The smallest ADC processing voltage is limited by the voltage of one least significant bit V_{LSB}. It is assumed here that the effective number of bits (ENOB) of the ADC is approximately the number of specified output bits. The voltage V_{LSB} is usually set to [15, 25]

$$V_{LSB} \leq A \cdot \min\{V_{BioSig,pp}\} \tag{2.4}$$

The required number of bits for the ADC is thus given by $\log_2(V_{DD}/V_{LSB})$ or equivalently by

$$ENOB \geq log_2\left(\frac{\max\{V_{BioSig,pp}\}}{\min\{V_{BioSig,pp}\}}\right) \quad (2.5)$$

with the input signal range[2] of the biomedical signal acquisition systems set to $\max\{V_{BioSig,pp}\}$.

Taking the ECG and EEG values from Table 2.2 as an example and using (2.5) would lead to a minimum ADC resolution of 11 bits for ECG and 7 bits for EEG applications. This calculation excludes the need for a higher number of bits due to abnormal high biomedical signals and measurement problems that are explained in the next section. Therefore, a higher number of bits is usually used to increase the input signal range. In the case of ECG and EEG, a minimum input signal range of approx. $10 - 20$ mVpp for ECG and $2 - 4$ mVpp for EEG is usually given. This corresponds to a minimum ADC resolution of 11-12 bits.

2.4.1 Problems associated with Signal Transduction and Input Leads

One of the common problems in the electrode/lead system are DC-offsets which originate mainly from the electrodes as explained in Section 2.3.1. These DC-signals are usually undesired except for some specific EEG measurements. Hence, a high-pass filter, removing any DC component represents an essential building block in most biomedical signal acquisition systems. A digital high-pass implementation to be applied after AD-conversion is generally possible. However, this would increase dramatically the requirements for the dynamic range of the ADC, hence an analog solution prior to the ADC is needed. In addition, high DC-offsets can lead to clipping or saturation within the amplifiers. The position of the high-pass within the analog chain therefore determines the systems sensitivity to clipping or saturation. A placement before any amplifier, i.e. AC-coupling (AC - alternating current) the system, would pose an optimal solution with respect to these problems. Nevertheless this is not always feasible, thence different interim solutions exists as explained below.

A further common problem are interferences due to power lines operating at a frequency of 50/60 Hz. These 50/60 Hz signals can couple capacitively into the measurement system which results in a superposition of the power line and biomedical signals at the acquisition output. The three main coupling mechanisms are (see also Fig. 2.7):

---

[2]The input signal range is also called ***input dynamic range*** for medical devices and is given in Vpp.

1. The power line signals couple over to the lead wires through $C_{pel,1}$ and $C_{pel,2}$. Assuming very high amplifier input impedance Z_{in}, currents $i_{pel,1}$ and $i_{pel,2}$ flow through the electrode impedances $Z_{el,1}$ and $Z_{el,2}$ to the patient grounding electrode. Any difference in electrode impedances will therefore create an input differential signal which will be amplified.

2. The power line signals couple over to the body via C_{pb} and a current i_{pb} flows through the body to earth via C_{be} (i_{be}) and to signal ground via the patient grounding electrode and Z_{Gnd}. This creates a common-mode voltage v_{cm} on the body with respect to ground.

3. The finite amplifier input impedance and the electrode impedances form a voltage divider. A common-mode voltage at the electrode inputs creates accordingly an differential signal at the input leads of the amplifier if there exists an imbalance between $Z_{el,1}$ and $Z_{el,2}$.

In order to reduce power line interferences, it becomes necessary to lower the electrode impedances to minimize their difference, this applies to 1. and 3. Input shielding the lead wires reduces furthermore the interference level for the first case. The mechanisms in 2. and 3. set two conditions to a differential amplifier used in biomedical application: First, the ability to reject common-mode voltages and second, a very high input impedance to lower the effect of the voltage division between electrode and amplifier impedance.

In addition, the electrodes exhibit noise (see Section 3.4 for an exact definition of noise terms) which depends on the electrode types used and should not exceed the input-referred noise level of the acquisition system by more than a factor of two [26]. Typical root-mean-square (RMS) noise values for standard electrodes lie in the range of a few μV up to hundreds of μV depending on electrode type and bandwidth considered [20, 26].

Another problem is the acquisition of additional biomedical signals that do not belong to the signal source under examination. An example for this would be the acquisition of EMG signals within ECG recordings or vice versa. These signals appear partly at both inputs of a differential amplifier, thus they can be considered as an additional common-mode voltage which would be attenuated by the rejection of common-mode signals of the differential amplifier. Additional error sources are given by movement artifacts and high-frequency interferences.

2.4.2 Biopotential (Pre-)Amplifier Requirements

Amplification within an front-end is typically performed by more than one gain stage with the first stage being a difference amplifier having a single-ended output. This amplifier (preamplifier) needs to have a high input impedance because of the relative high biomedical signal source impedance and to suppress interferences. Therefore, an instrumentation amplifier (IA) having an input impedance of $10 - 100$ MΩ is used for the first stage.

Common Mode Rejection

The ability of the IA to amplify only differential signals is limited by non-idealities like mismatch between gain setting resistors. The need to suppress common-mode signals has been outlined above and is given qualitative by the common-mode rejection ratio (CMRR) of the IA:

$$\mathrm{CMRR} = \frac{A_d}{A_{cm}} \tag{2.6}$$

with A_d as the differential-mode gain, A_{cm} as the common-mode gain and CMRR usually given in dB. The CMRR value at 50/60 Hz is of particular importance because of power line interferences. Typically, a CMRR value of more than 80 dB is required for biomedical acquisition systems [13]. The CMRR of the biomedical amplifier system can be increased using an active patient grounding also known as driven right leg (DRL) circuit [13] and shown in Fig. 2.8. The common-mode voltage on the body of the patient, V_{cm}, is sensed using two resistors R_a, inverted, amplified, and finally fed back to the ground electrode. An operational amplifier is used in this circuit as an inverting amplifier and provides a gain of $-2R_f/R_a$. Resistors R_0 and R_{fb} limit the output current and should have values in the megaohm range. The common-mode voltage and hence the CMRR of the biomedical amplifier system is accordingly attenuated by the action of the inverting amplifier.

Noise and Linearity

The amplifier noise should be low enough in order to have no influence on the minimal signal-to-noise ratio of the signal acquisition system. An amplifier noise level far beyond this constraint has no benefit and unnecessarily increases the power dissipation of the system. It is therefore advantageous to choose the input-referred amplifier RMS noise value somewhat smaller than the RMS quantization noise of the ADC which is $V_{LSB}/\sqrt{12}$, thus

$$v_{ni} \leq \frac{V_{LSB}}{G\sqrt{12}} \tag{2.7}$$

When signal averaging is performed, like in the case of Evoked Potentials, where the summing of n recordings leads to a SNR improvement by a factor of $\sqrt{(n)}$, a higher amplifier noise level is clearly acceptable.

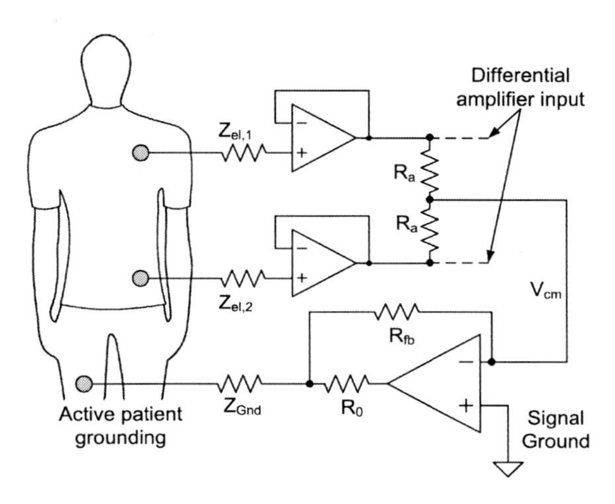

Figure 2.8: Driven right leg system.

Nonlinearities within amplification introduce distortions in the output signal. This applies particularly to signals exceeding the input and output voltage ranges of the amplifier. The linearity of the amplifier is specified by its total harmonic distortion (THD) which is given by

$$\mathrm{THD} = \sqrt{\frac{a_2^2 + a_3^2 + a_4^2 + \ldots + a_n^2}{a_1^2}} \tag{2.8}$$

with a_1 being the input tone amplitude and $a_2 \ldots a_n$ being its first n harmonics. Usually, a THD of 0.1-1% is specified for voltages within the dynamic range of the biomedical amplifier.

Power, Gain and Additional Parameters

The AFE power consumption can be a limiting factor for multiple channel systems being battery operated. An amplifier consuming less power exhibits normally also a higher noise, giving rise to a noise-power trade-off. In order to compare different biomedical amplifiers with respect to their noise-power performance a figure of merit called *noise efficiency factor* NEF is used, it is defined as [27]

$$\mathrm{NEF} = v_{ni}\sqrt{\frac{2\,I_{tot}}{\pi\,U_T\,4k_BT\,BW}} \tag{2.9}$$

where I_{tot} is the current consumption of the amplifier, U_T is the thermal voltage, k is the Boltzmann constant, T is the absolute temperature and BW is the bandwidth of the amplifier. The NEF compares the noise-power performance of the amplifier to that of an ideal bipolar transistor with NEF=1.

The gain of the amplifier should be matched to the ADC to exploit its full resolution. Also, its accuracy must meet the given specifications. Different gain error definitions are used in biomedical acquisition systems, including absolute or inter-channel gain errors and gain accuracy referenced to one LSB value. Exemplary values are an absolute gain error of ± 5 % for ECG devices [28] and an inter-channel gain error of < 1 % for the biomedical system solutions that are described in Sections 5.2 and 5.3. The option of a selectable gain is furthermore advantageous to allow the physician to change the input dynamic range.

Two additional parameters, cross-talk and isolation, are closely related to a single amplifier. Cross-talk describes signal coupling from one channel to the others and becomes particularly important in integrated solutions where channels are placed next to each other within a small distance. It is defined as

$$\mathrm{XT}_{Ch} = \frac{S_i}{S_j} \quad (i, j = 1 \ldots n \text{ and } i \neq j) \tag{2.10}$$

with S_i being the output signal power of one channel having a test signal at its input and S_j denotes the output signal power of all other channels which are grounded at their inputs. The cross-talk ratio S_i/S_j is normally given in dB and should stay above a value of 60 dB or more for all channel pairs.

In contrast to battery operated systems, an isolation barrier is needed between the input circuitry that is connected to the patient and the mains power. The isolation can be situated in the analog part of the front-end by using an isolation-amplifier. This introduces an additional amplifier parameter called *isolation mode rejection ratio* (IMRR). The IMRR measures the quality of isolation and is defined as the ratio between signal gain and isolation voltage gain v_{im} where v_{im} refers to the voltage across the isolation barrier. The isolation barrier can also be placed within the digital domain of the biomedical acquisition system, e.g. by using opto-couplers for digital output data lines.

Bandpass and DC-Suppression

In order to realize the bandpass characteristic of the biomedical signal acquisition chain both high- and low-pass filters are needed. Low-pass filtering is used to remove noise outside of the specified bandwidth and to prevent anti-aliasing, hence it is placed just before the ADC. The high pass filter is realized either by AC-coupling the preamplifier or by a high-pass filter following a DC-coupled preamplifier; Table 2.3 shows the advantages and disadvantages of both solutions.

Parameter/Effect	AC-coupling	DC-coupling
Coupling capacitor charging	Only for BJT (base current)	n.a.
Very low high-pass cut-off frequency	Large resistors needed	Ideal
CMRR	Reduced	Optimal
Response to high electrode DC-offsets	In principle ideal	Saturation of the preamplifier

Table 2.3: Comparison of AC- or DC-coupling the preamplifier [13, 14].

There exists yet another way of removing electrode induced DC-offset at the input of the amplifier which is called active AC-coupling or DC-suppression [29], Fig. 2.9(a) shows the working principle for an analog solution and Fig.(b) of a mixed-signal version.

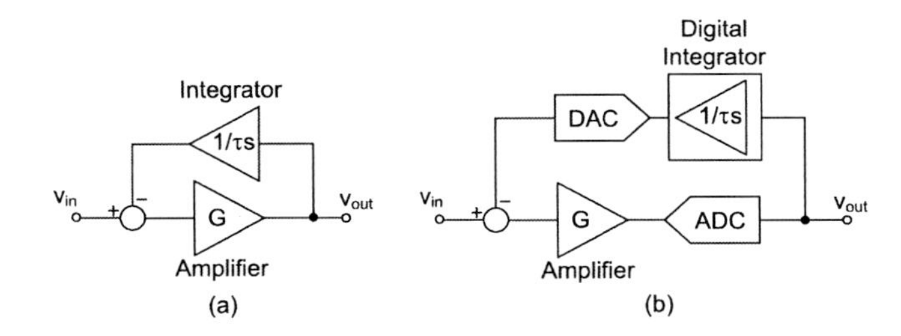

Figure 2.9: Active DC-suppression.(a) All analog; (b) Mixed-signal.

The working principle is based on a feedback loop to subtract the DC-offset at the input terminal of the amplifier. The feedback signal is generated by feeding the output signal of the amplifier to an integrator and using its output signal. The systems transfer function becomes accordingly

$$H(s) = \frac{A\tau s}{A + \tau s} \tag{2.11}$$

with τ being the time constant of the integrator. This corresponds to a high-pass transfer function with cut-off frequency $\omega_c = A/\tau$.

Polarity and Signal Reference

The use of a difference amplifier like the IA allows to measure a *true* voltage difference independently of the signal ground. However, some biomedical signal acquisitions require to reference the input signal against a common point on the body of the patient. This leads to a differentiation in electrode and amplifier configuration, namely *bipolar* and *unipolar* electrode arrangements, Fig. 2.10 shows both the unipolar and two bipolar setups.

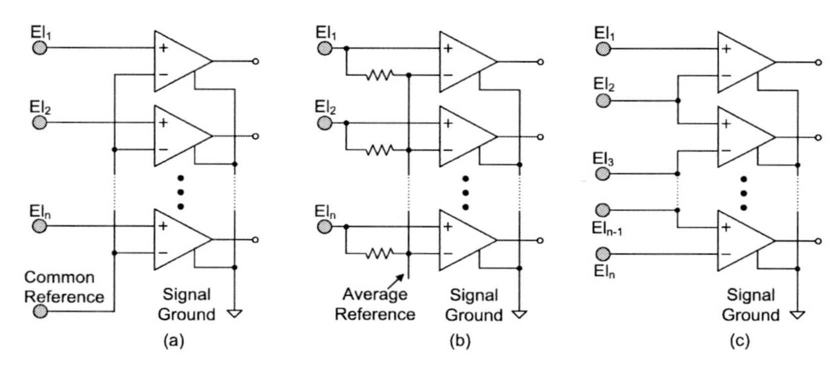

Figure 2.10: Polarity and Signal Reference. (a) Unipolar with common reference. (b) Unipolar with average reference. (c) Bipolar.

For the bipolar arrangement the difference between two electrodes is measured whereas for unipolar a reference signal is needed. This reference signal can be obtained using a separate reference electrode or by averaging the input signals. The simplest way to accomplish the averaging is using 1 MΩ resistors connected from each electrode to a single reference line whereas modern systems use digitally assisted averaging [10]. The setups shown apply mainly to EEG measurements [6], the principle of bipolar or unipolar electrodes is nevertheless used, sometimes in a slightly modified way, for all other biomedical signal types.

2.5 Three Generic Types of Analog Front-Ends

Based on the descriptions given, three main front-end topologies can be identified. All three include the band-pass characteristics, an instrumentation amplifier (IA) used as a preamplifier and a programmable gain amplifier (PGA) which is also referred to as postamplifier. The topologies differ mainly in the strategy to implement a high-pass behavior in order to eliminate electrode DC-offsets. There exists also mixtures of these basic types, e.g. by using a DC-coupled preamplifier and a DC-suppression circuitry which has been implemented in the system solutions, see Sections 5.2 and 5.3.

2.5.1 Type A – Gain-HP-gain

This front-end uses a DC-coupled IA followed by a high-pass filter, its block diagram is shown in Fig. 2.11. The advantages of this solution are a relative high CMRR, a high input impedance and the possibility to measure directly the DC-potential when bypassing the high-pass. The main drawback of this topology is its tendency to saturate at the preamplifier for larger DC-offsets, hence a low amplification factor is needed for the instrumentation amplifier. This front-end solution represents more or less a *standard front-end* described in many textbooks [13, 14] and has been implemented in integrated solutions [25, 30].

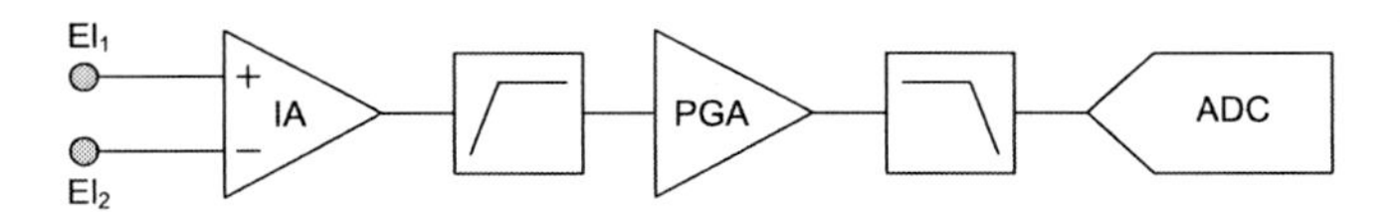

Figure 2.11: Block diagram of the Gain-HP-gain type front-end.

2.5.2 Type B – AC-coupled

The AC-coupled front-end depicted in Fig. 2.12 consists of an AC-coupled IA and is used in measurements where a high electrode DC-offset is expected. This applies mainly to implanted systems using platinum or gold coated electrodes which exhibit very high DC-offsets, an example system is described in [31]. In these systems the ability to tolerate high DC-offsets is traded in for a lower CMRR and a more complex input circuitry.

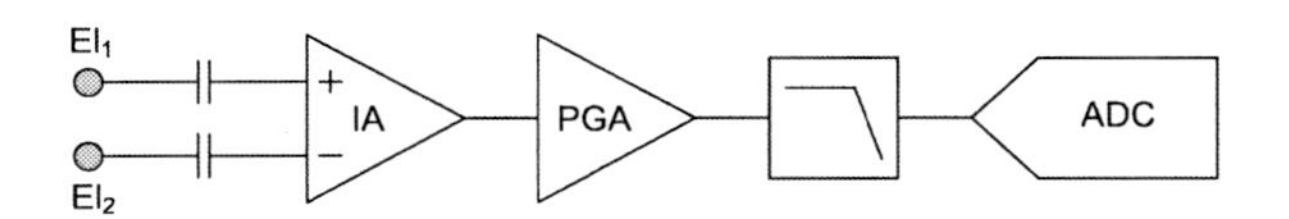

Figure 2.12: Block diagram of the AC-coupled type front-end.

2.5.3 Type C – DC-Suppression

The use of an active DC-suppression allows to exploit advantages of both the AC- and DC-coupled systems which results in a very high CMRR and the ability to suppress large DC-voltages. Its drawback is a rather complex circuit implementation and the possibility of being instable because of the feedback used. The DC-suppression circuit can be implemented either analog or digitally, Fig. 2.13 shows a system using the analog version. Biomedical signal acquisition circuits using analog circuitry are presented in [29, 32] whereas a digital version has been implemented for the systems described in Chapter 5.

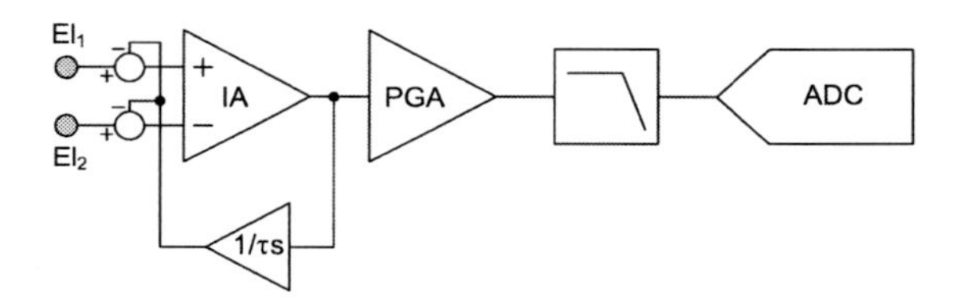

Figure 2.13: Block diagram of the DC-suppression type front-end.

Chapter 3

Analog CMOS

Using modern CMOS (complementary metal-oxide-semiconductor) technology to design biomedical signal acquisition systems poses both a challenge and a facilitation with respect to analog and digital design, respectively. The continuous downscaling of the MOSFET (metal-oxide-semiconductor field-effect transistor) provides a basis for digital circuits having a high functionality due to a high device count in conjunction with lower power consumption of each device. The increasing number of physical effects to be considered in analog design diminishes the benefits of downscaling in the analog domain. Nevertheless, the necessity to use modern CMOS technology for analog design is given by the demand to integrate both, digital and analog circuits into one integrated circuit called system-on-chip (SoC), thereby reducing system cost. At present, there is a wide range of CMOS technology nodes available starting at approximately 0.8 μm, which is used mainly for analog circuits, down to 45 nm for state-of-the-art digital systems. Within this range an appropriate decision has to be taken for the biomedical system to be designed.

Main characteristics of CMOS technology scaling are given in Section 3.1 and Section 3.2 details MOSFET operating points. Section 3.3 describes the resulting small-signal models and Section 3.4 gives an overview of MOSFET noise. Finally, additional short channel effects are discussed in Section 3.5.

3.1 MOSFET Scaling

In general, scaling or shrinking of MOSFET devices should keep their characteristics or even improve some as described by Dennard in 1974 [33]. To achieve this, the electric fields inside the device have to stay constant if geometric dimensions are scaled down by a factor k_s. A set of rules, shown in Table 3.1, describes how to change the device parameters in accordance to the *constant field scaling theory*.

Device Parameter	Factor
Device dimension t_{ox}, L, W, x_j	$1/k_s$
Doping concentration N_a	k_s
Voltage, Current V, I	$1/k_s$
Capacitance C	$1/k_s$
Delay time t_d	$1/k_s$
Power dissipation P	$1/k_s^2$
Power density P/A	1

Table 3.1: Constant field scaling rules [33].

Application of the above rules is limited by approaching smaller MOSFET geometries as physical effects that could be neglected before becoming more pronounced. In addition, not all results of the scaling rules are advantageous for the design of analog circuits such as V_{DD} scaling. The list of physical effects that impede or hinder the downscaling of MOSFETs include [34]

- Threshold voltage V_{TH} scaling limited by increased sub-threshold leakage
- Reduced mobility and increased junction leakage due to high channel doping
- Short channel effects (SCE) like threshold roll-off
- Gate current leakage as a result of ultra-thin oxides

Additional problems, specifically important in analog design, are an increasing device mismatch and lower supply voltages [35, 36, 37] with both having an impact on the dynamic range (DR) for signal processing.

A dynamic range which depends on the supply voltage V_{DD} is given by the proportionality

$$\mathrm{DR}_{V_{DD}} \propto \frac{V_{DD}^2}{v_n^2} - \mathrm{SNR}_{min} \tag{3.1}$$

with v_n as the total RMS noise and SNR_{min} is the minimum signal-to-noise ratio (SNR).

In addition, a dynamic range linked to the mismatch of the transistors used can be given [36]. The associated dynamic range DR_{mis} is now defined as the ratio of the signal voltage V_{sig} and mismatch voltage V_{mis}. The mismatch voltage V_{mis} is mainly related to the mismatch in threshold voltage for modern CMOS technologies, thus $V_{TH} \approx 6\,\sigma(\Delta V_{TH})$, with $\sigma(\Delta V_{TH})$ being the standard deviation of V_{TH} (see also Section 3.5.1). With scaling, parameter $\sigma(\Delta V_{TH})$ increases and hence DR_{mis} decreases.

The constant field scaling rules apply also to the supply voltage V_{DD} which has been reduced accordingly with scaling. However, the magnitude of V_{DD} is closely related to the magnitude of the threshold voltage V_{TH} in order to maintain signal integrity within the digital domain. The slow-down in V_{TH} reduction due to sub-threshold leakage implies therefore a slow-down in V_{DD} reduction. For older technologies, a supply voltage of $V_{DD} = 5$ V was common. Starting with 350 nm, V_{DD} has been constantly reduced and seems to level off around 1 V for newer technologies, Fig. 3.1 shows the scaling of V_{DD} with respect to technology nodes [37] and the resulting dynamic range reduction.

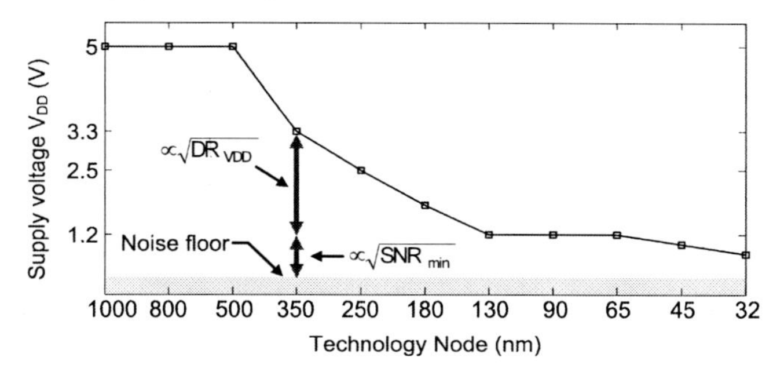

Figure 3.1: Scaling of V_{DD} [37] and reduction of $\mathrm{DR}_{V_{DD}}$ with respect to technology nodes (the noise floor is not to scale). SNR_{min} indicates the minimal SNR for an analog signal processing circuit.

The influence of a lower supply voltage on the power consumption P_{diss} and dynamic range $\mathrm{DR}_{V_{DD}}$ can be demonstrated at a class-B output stage of an (ideal) amplifier driving a load resistor R_L [38], see Fig. 3.2. The voltage across R_L is a sine wave with a peak voltage of $V_{DD}/2$. In class-B mode, the top transistor in Fig. 3.2 conducts a current I_p for the positive half wave with the bottom transistor turned off, whereas for the negative half wave the top transistor turn offs and the bottom transistor conducts a current I_n. The application of the output signal voltage in class-B mode on R_L having a thermal voltage noise of $v_{n,R} = \sqrt{4kTR_L BW}$ results in a maximum dynamic range of

$$\mathrm{DR}_{V_{DD},max} = \frac{V_{DD}^2}{32kTR_L}\frac{1}{BW} \tag{3.2}$$

and using (3.2) in a dynamic range to supply power ratio of

$$\frac{\mathrm{DR}_{V_{DD},max}}{P} = \frac{\pi}{16kT}\frac{1}{BW} \tag{3.3}$$

where P is the supply power dissipation. The dynamic range is further reduced by taking into account amplifier non-idealities like noise, offset and mismatch [39]. The consequence arising from (3.3) is that the power consumption will not decrease for lower supply voltages in order to keep the dynamic range constant.

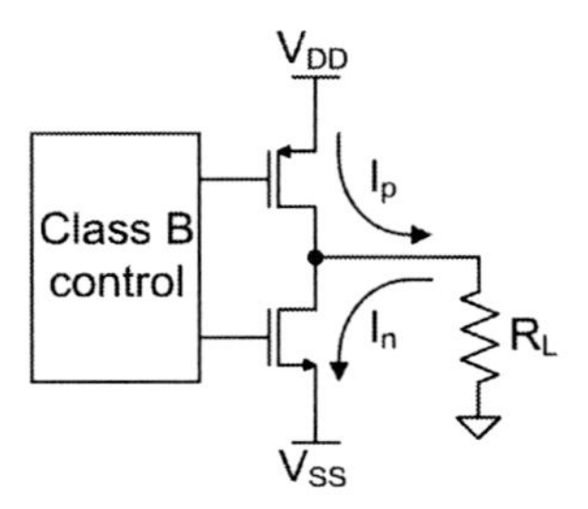

Figure 3.2: Class-B output stage driving a signal processing resistor R_L [38].

The increased amount of physical effects to be considered in nanometer CMOS technology implies also an increase in the complexity of MOSFET modeling. This adds a difficulty for analog design methodology, which normally begins with hand calculations of the circuit to be designed. Whereas basic SPICE (Simulation program with integrated circuit emphasis) level models were quite well suited for performing hand calculation with older technologies, the currently used BSIM (Berkeley Short-channel IGFET Model) models impede hand calculations as a result of a high parameter number; Fig. 3.3 shows the evolution of design parameter count with continuous reduction of channel length [40]. Additionally, the increasing complexity of equations makes it difficult to predict how MOSFET characteristics change in response to design parameter changes, e.g. by increasing the device length.

One approach to solve this problem is the use of basic approximation for hand calculation to obtain a first set of design parameters and then arrive at the final solution with the aid of simulation. This approach has been used within the scope of this work. The approximated parameters used have to be extracted from BSIM parameters. An alternative approach, is the gm/I_D methodology described in [41].

3.2 MOSFET Operating Point

The design of analog circuits using MOS transistors requires a profound knowledge of its characteristics and physics, on the one hand to exploit certain physical effects for circuit functionality and, on the other hand, to avoid those unwanted. A simplified view of an n-channel MOSFET (the substrate is p-type doped and the two n^+

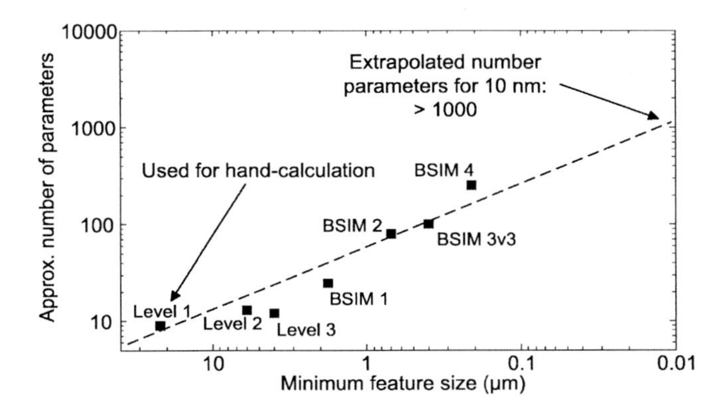

Figure 3.3: Number of SPICE/BSIM parameters versus structure size.

regions form the source and drain) including several parameters needed to derive its current-voltage (I-V) characteristics is shown in Fig. 3.4.

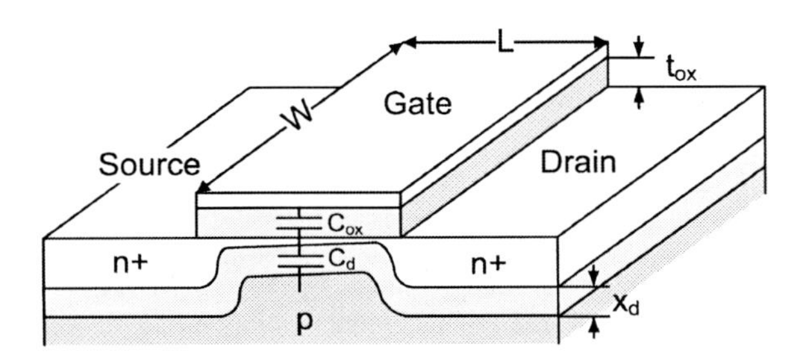

Figure 3.4: Structure of a n-channel MOSFET (NMOS).

The following derivations of MOSFET characteristics are accomplished for a n-channel (NMOS) device with threshold voltage V_{THn}. For p-channel devices (PMOS) with threshold voltages V_{THn}, the same equations can be used by reversing the polarities of all currents and voltages (assuming that in this case the drain current I_D flows from drain to source).

3.2.1 Basic MOSFET Parameters

Fig. 3.4 shows two capacitances within the MOS structure with $W \cdot L \cdot C'_{ox}$ being the capacitance across the gate oxide and $W \cdot L \cdot C'_d$ is the capacitance related to the depletion layer. The capacitance normalized to unit area values for C'_{ox} and C'_d can be calculated using

$$C'_{ox} = \frac{\varepsilon_{ox}}{t_{ox}} \text{ and } C'_d = \frac{\varepsilon_{si}}{x_d} \tag{3.4}$$

where t_{ox} is the oxide thickness, x_d is the width of the depletion region, and ε_{ox} and ε_{si} are the SiO_2 and Si dielectric constants, respectively. The width x_d is depending on the voltage across the depletion region, i.e. it is changed by the source-bulk voltage (V_{SB}). The capacitance ratio C'_{ox}/C'_d influences the MOSFETs' subthreshold behavior, threshold voltage and hence the body effect. Therefore, a parameter n which depends on this ratio is defined for device modeling:

$$n = 1 + \frac{C'_d}{C'_{ox}} \tag{3.5}$$

Typical values for n lie in the range of 1 to 1.5 [42].

In addition, a source-bulk voltage dependent parameter γ called *body effect parameter* is usually chosen to model the change in threshold voltage V_{THn} from a zero source-bulk voltage bias threshold voltage V_{T0}:

$$V_{THn} = V_{T0} + \gamma \left(\sqrt{|2\phi_F| + V_{SB}} - \sqrt{|2\phi_F|} \right) \tag{3.6}$$

with ϕ_F being the Fermi potential and γ defined as

$$\gamma = \frac{\sqrt{2\varepsilon_{si} q N_A}}{C'_{ox}} \tag{3.7}$$

where N_A is the doping concentration.

Another basic parameter is the channel charge carrier mobility. The bulk or low-field surface mobility μ_0 differs from the measured mobility in the presence of a high vertical electrical field resulting from the gate voltage [43]. The deviation from μ_0 is attributed to Coulombic scattering at ions within the interface and to surface roughness scattering. The resulting effective mobility μ_{eff} depends on the effective normal field E_{eff} and can be expressed as

$$\mu_{eff} = \frac{\mu_0}{1 + (E_{eff}/E_0)^{\nu}} \tag{3.8}$$

with values for the parameters E_0, μ_0 and ν as given in Table 3.2. The effective normal field E_{eff} can be approximated by the average effective field using [44]

$$E_{eff} \approx \frac{V_{GS} - V_{THn}}{6t_{ox}} \tag{3.9}$$

Parameter	Electron (surface)	Hole (surface)
$\mu_0\,(\mathrm{cm}^2/\mathrm{Vs})$	670	160
$E_0\,(\mathrm{MV/cm})$	0.67	0.7
ν	1.6	1

Table 3.2: Parameters for the effective mobility [44].

The validity of approximation is limited to n^+-polysilicon n-channel MOSFETs and to p^+-polysilicon p-channel MOSFETs.

An additional effect to be considered in the derivation of the I-V characteristics is the difference between the geometrical mask dimensions of the gate and the actual or electrical geometries. The effective gate length L_{eff} differs from the drawn gate length L_{drawn} as a result of underdiffusion. The amount of underdiffusion is a function of drain and source junction depth and is given by the parameter ΔL, thus, $L_{eff} = L_{drawn} - 2\Delta L$. The effective channel width W_{eff} can be defined in a similar way, hence $W_{eff} = W_{drawn} - 2\Delta W$[1].

3.2.2 MOSFET I-V characteristics

An exemplary NMOS device output characteristic is shown in Fig. 3.5a. It can be subdivided into different operating regions with respect to the applied voltages V_{DS} and V_{GS} as illustrated in Fig. 3.5b [45]. For a constant V_{GS}, V_{DS} sets the MOSFET to be either in *linear (triode)* or *saturation* region and for a constant V_{DS}, V_{GS} classifies the I-V curve into three distinct regions within the saturation region. These three regions are *strong inversion (si)*, which is the region modeled by the classic *square-law* equation, *weak inversion (wi)*, where the drain current is governed by diffusion current and *velocity saturation (vs)*, in which the channel charge carriers reach their maximum velocity.

When increasing the gate-source voltage V_{GS} for a MOSFET in weak inversion and biased with a high V_{DS} (saturation), the MOSFET first enters the strong inversion region at the gate-source voltage $(V_{GS})_{ws}$. It traverses through this region for an increasing V_{GS} and enters finally the velocity saturation region at the gate-source voltage $(V_{GS})_{sv}$.

The vertical size of the strong inversion region in Fig. 3.5b depends on the channel length, for short channel devices it becomes very small which can be seen from (3.26) in Section 3.2.6 below.

[1]The following convention is used in this work if the channel length or width are not explicitly labeled as effective or drawn: L_{eff}, W_{eff} are implied in the context of equations and L_{drawn}, W_{drawn} in the context of design parameters.

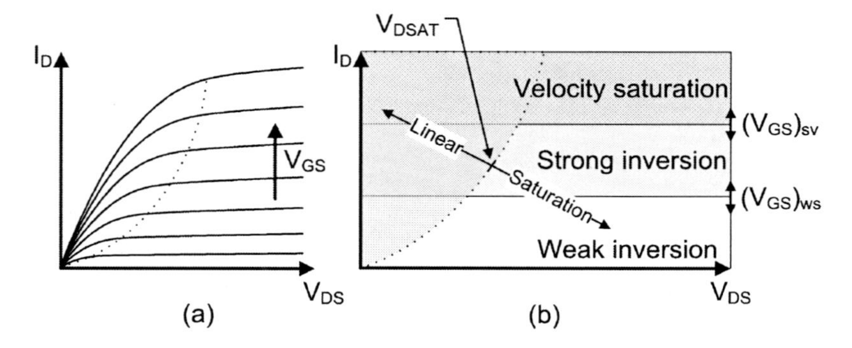

Figure 3.5: MOSFET operating regions. (a) Exemplary NMOS I_D versus V_{DS} characteristic. (b) Schematic diagram of the operating regions within an NMOS I_D versus V_{DS} characteristic - the regions are not to scale for illustration purposes.

3.2.3 Linear Region

For the transition voltage between the linear and saturation region, $V_{DS} = V_{DSAT}$ is defined by the condition that $dI_D/dV_{DS} = 0$ for the linear region drain current when neglecting channel length modulation and velocity saturation (see Section 3.2.4 and 3.2.6 for a description of these effects). The MOSFET enters the linear region for $V_{DS} < V_{DSAT}$ and its drain current is expressed as:

$$I_D = \mu_{eff,n} C'_{ox} \frac{W}{L} \left(V_{GS} - V_{THn} - \frac{nV_{DS}}{2} \right) V_{DS} \tag{3.10}$$

The factor n has been included to model the effect of bulk-charge which effectively reduces I_D. However, with n having values of 1 to 1.5, this factor is often neglected for basic calculation, i.e. it is assumed that $n \approx 1$. The transition voltage between the linear and saturation region can now be calculated as

$$V_{DSAT}{}^* = V_{OV} = (V_{GS} - V_{THn})/n \approx (V_{GS} - V_{THn}) \tag{3.11}$$

where asterisk in V_{DSAT} indicates that this value is only valid for long channel MOSFETs. Its value corresponds (for long channel MOSFETs) to the *overdrive voltage* V_{OV}, i.e. the gate voltage applied in addition to the threshold voltage needed to inverted the channel ($V_{GS} = V_{THn}$).

3.2.4 Strong Inversion (Saturation)

The MOSFET enters the saturation region for $V_{DS} > V_{DSAT}{}^{*}$. In strong inversion $(V_{GS})_{sv} > V_{GS} > (V_{GS})_{ws}$ and the drain saturation current is (square-law)

$$I_D = \frac{KP_n}{2}\frac{W}{L}(V_{GS} - V_{THn})^2 \tag{3.12}$$

with KP_n being the current factor which is defined as

$$KP_n = \frac{\mu_{eff,n} C'_{ox}}{n} \approx \mu_{0,n} C'_{ox} \tag{3.13}$$

A current factor KP_p is defined in a similar way for PMOS transistors. In addition to neglecting the effect of bulk charges in the approximation for KP_n in (3.13), μ_{eff} has been replaced by μ_0 which is only valid for low gate-source voltages, see (3.8) and (3.9).

Equation (3.12) includes no V_{DS} dependence of the drain current in saturation or, as seen from the analog design view, the small-signal output resistance r_{ds} or output conductance $g_{ds} = 1/r_{ds}$, respectively. There are three mechanisms giving raise to the MOS output resistance, namely *channel length modulation (CLM)*, *drain induced barrier lowering (DIBL)* and *substrate current induced body effect (SCBE)*. For lower V_{DS}, CLM is the main factor causing output resistance, whereas DIBL is the dominant at medium and SCBE at higher drain-source voltages [46]. The dependence of the drain current on V_{DS} can be modeled in (3.12) using a factor λ or its inverse V_A resembling the Early voltage of bipolar transistors:

$$\lambda = \frac{1}{V_A} = \frac{1}{V_{ACLM}} + \frac{1}{V_{ADIBL}} + \frac{1}{V_{ASCBE}} \tag{3.14}$$

The parameters V_{ACLM}, V_{ADIBL} and V_{ASCBE} describe the influence of each effects to V_A (or λ). Their values depend in general on several MOSFET parameters, most importantly on the channel length L. Therefore, V_A is often approximated by a constant value V'_A given in (V/μm) which is scaled by the channel length L given in (μm) [47]:

$$\lambda \approx \frac{1}{V'_A L} \tag{3.15}$$

Hence, a modified version of (3.12), including the V_{DS} dependence can be formulated as

$$I_D = \frac{KP_n}{2}\frac{W}{L}(V_{GS} - V_{THn})^2 (1 + \lambda V_{DS}) \tag{3.16}$$

3.2.5 Weak Inversion

For gate-source voltages $V_{GS} < (V_{GS})_{ws} \approx V_{THn}$, i.e. in the weak inversion region, I_D starts to decrease exponentially with respect to V_{GS}. The charge transport mechanism between drain and source in this region is dominated by diffusion current in contrast to drift current in strong inversion. Therefore, the MOSFET resembles a bipolar junction transistor (BJT) in this region and the drain current becomes (assuming $V_{SB} = 0$) [48]

$$I_D = I_{D0}\frac{W}{L}e^{(V_{GS}-V_{THn})/nU_T}\left(1 - e^{-V_{DS}/U_T}\right) \tag{3.17}$$

with I_{D0} being the normalized specific current, $(kT/q) \approx 26$mV at 300 K, and n as defined in (3.5). In this context, n is referred to as the *subtreshold slope factor*. The normalized specific current I_{D0} is defined as

$$I_{D0} = 2n\mu_0 C'_{ox}U_T^2 \tag{3.18}$$

Both saturation and linear region are modeled using (3.17). The term $(1-e^{-V_{DS}/U_T})$ in (3.17) becomes negligible for a V_{DS} larger than a few U_T's which defines approximately the point of transition between linear and saturation region. As in the case of strong inversion, an output resistance is present within saturation. This can be likewise modeled using a parameter λ:

$$I_D = I_{D0}\frac{W}{L}e^{(V_{GS}-V_{THn})/nU_T}\left(1 + \lambda V_{DS}\right) \tag{3.19}$$

The gate-source voltage V_{GS} separating weak from strong inversion $(V_{GS})_{ws}$ is derived by equating (3.16), (3.19) and their first derivatives, giving [47]

$$(V_{GS})_{ws} = 2nU_T + V_{THn} \tag{3.20}$$

From this it follows that the transition voltage lies approximately $65 - 80$ mV above the threshold voltage V_{THn}. The MOSFET is often operated in strong inversion near the onset of weak inversion to obtain an optimum between speed and current efficiency. To keep a safety margin from going into weak inversion, the overdrive voltage V_{OV} is set to a value of 2-3 times $((V_{GS})_{ws} - V_{THn})$, i.e. $V_{OV} \approx 200$ mV. This value is relatively independent of technology scaling as indicated by (3.20).

When a MOSFET is driven from weak to strong inversion, it passes through a transition region called *moderate inversion*. A way to model this transition was proposed by Enz, Krummenbacher and Vittoz in their analytical MOS model (EKV model) [49] using an interpolation function:

$$F(x) = \left[\ln\left(1 + e^{x/2}\right)\right]^2 = \begin{cases} \left(\frac{x}{2}\right)^2 & \text{for } x \gg 0 \\ e^x & \text{for } x \ll 0 \end{cases} \tag{3.21}$$

In addition, all MOSFET electrode voltages (i.e. V_G, V_D and V_S) are referenced to the bulk potential in contrast to the usual SPICE/BSIM convention using the source electrode. The MOSFET is described completely symmetric for all regions using a *forward current* I_F referenced to V_S and a *reverse current* I_R referenced to V_D with the drain current given as $I_D = I_F - I_R$ (see [49] for their definition). Using (3.21) the model can be expressed as source referenced by assuming saturation and V_{THn} as defined in (3.6) [42]

$$I_D = \frac{KP_n}{2}\frac{W}{L}\left(V_{GS} - V_{THn}\right)^2 \left[ln\left(1 + e^{(V_{GS}-V_{THn})/2nU_T}\right)\right]^2 \tag{3.22}$$

whereby the desired (smooth) transition is obtained.

3.2.6 Velocity Saturation

The charge carrier velocity saturates at very high electric fields, hence the name *velocity saturation* for the MOSFET region dominated by this effect. The electric field at the onset of velocity saturation is labeled as E_{sat} and the corresponding maximal carrier velocity ν_{sat} is approximately $10^7 cm/s$ for both electrons and holes. As channel lengths are scaled down this becomes more pronounced due to increasing electric fields between source and drain.

Because of velocity saturation the drain current is no longer a quadratic function, but depends linearly on V_{GS} for high currents (i.e. large V_{GS}), thus

$$I_D = WC'_{ox}\nu_{sat}\left(V_{GS} - V_{THn}\right) \tag{3.23}$$

If V_{GS} is reduced, the MOS transistor enters again strong inversion at $\left(V_{GS}\right)_{vs}$, however, $\left(V_{GS}\right)_{vs}$ approaches more and more V_{THn} for smaller technologies [50]. This transition gate voltage can be derived by equating the first derivatives of (3.12) and (3.23) with respect to V_{GS} [45], which gives

$$\left(V_{GS}\right)_{vs} = 2\frac{\nu_{sat}}{\mu_{eff}}L + V_{THn} = E_{sat}L + V_{THn} \tag{3.24}$$

where

$$E_{sat} = \frac{2\nu_{sat}}{\mu_{eff}} \tag{3.25}$$

As in the case of the transition between weak and strong inversion, a model including both strong inversion and velocity saturation is needed. Including the effect of velocity saturation into the derivation of the square-law model for the drain current results in [43]

$$
\begin{aligned}
I_D &= \frac{\mu_{eff} C'_{ox}}{2} \frac{W}{L} \frac{(V_{GS} - V_{THn})^2}{1 + \dfrac{(V_{GS} - V_{THn})}{E_{sat} L}} \\
&= \begin{cases} W C'_{ox} \nu_{sat} (V_{GS} - V_{THn}) & \text{for } E_{sat} L \ll (V_{GS} - V_{THn}) \\ \frac{KP_n}{2} \frac{W}{L} (V_{GS} - V_{THn})^2 & \text{for } E_{sat} L \gg (V_{GS} - V_{THn}) \end{cases}
\end{aligned} \tag{3.26}
$$

The saturation drain voltage V_{DSAT} is accordingly smaller than the long channel drain voltage V_{DSAT}^* or the overdrive voltage V_{OV}. The exact value of V_{DSAT} is calculated using [43]

$$
V_{DSAT} = \frac{E_{sat} L (V_{GS} - V_{THn})}{E_{sat} L + (V_{GS} - V_{THn})} \leq V_{DSAT}^* = V_{OV} \tag{3.27}
$$

3.2.7 Equation Overview

In order to summarize the MOSFET I-V characteristics presented in the above sections, Table 3.3 gives an overview of the equations for the drain current I_D depending on the region of operation.

<table>
<tr><th>Saturation Region $(V_{DS} > V_{DSAT}^{(*)})$</th><th>$V_{GS}$</th><th colspan="3">Equations for I_D</th></tr>
<tr><td>Velocity saturation</td><td>↑
$(V_{GS})_{sv}$</td><td>(3.23)</td><td></td><td rowspan="2">(3.26)</td></tr>
<tr><td>Strong inversion</td><td>↑
$(V_{GS})_{ws}$</td><td>(3.12), (3.16)</td><td rowspan="2">(3.22)</td></tr>
<tr><td>Weak inversion</td><td>↑</td><td>(3.17), (3.19)</td><td></td></tr>
<tr><td>Linear Region $(V_{DS} < V_{DSAT}^{(*)})$</td><td></td><td colspan="3">(3.10)</td></tr>
</table>

Table 3.3: Overview of the equations for the drain current depending on the region of operation.

3.3 Small-Signal Parameters

The parameters given in this section apply for both, weak and strong inversion and the MOSFET is assumed to be outside moderate inversion. A low to medium frequency MOS transistor small-signal model is depicted in Fig. 3.6.

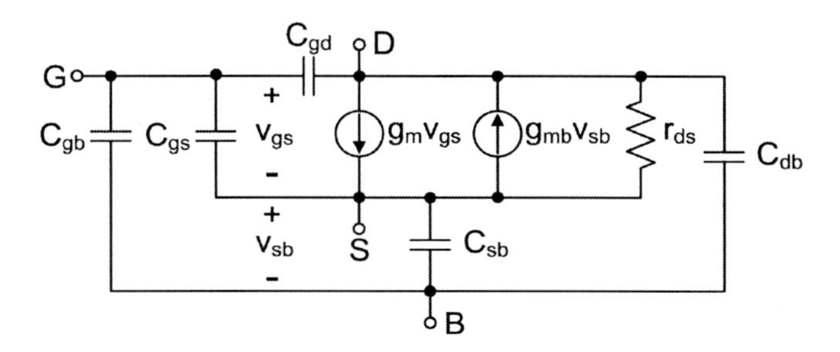

Figure 3.6: Small-signal equivalent circuit.

3.3.1 Transconductance and Output Resistance

The transconductance of MOS transistors is defined as the first derivative of the drain current I_D with respect the the gate-source voltage V_{GS}. In strong inversion the drain current is given by (neglecting channel length modulation) (3.12), thus

$$g_m = \frac{\partial I_D}{\partial V_{GS}} = \sqrt{2KP_n \frac{W}{L} I_D} = \frac{2I_D}{V_{GS} - V_{THn}} = KP_n \frac{W}{L} \left(V_{GS} - V_{THn}\right) \tag{3.28}$$

where the g_m has been expressed differently with respect to the design parameters (W/L), V_{GS} and I_D.

The g_m in weak inversion is obtained using (3.19) and neglecting channel length modulation, thus

$$g_m = \frac{I_D}{nU_T} \tag{3.29}$$

This transconductance resembles that of a BJT with the exception of the additional parameter n. The transconductance for velocity saturation is derived from (3.23), which gives

$$g_m = WC'_{ox}\nu_{sat} \tag{3.30}$$

The approach to model the output resistance using λ is identically for weak and strong inversion. Accordingly,

$$g_{ds} = \frac{1}{r_{ds}} = \frac{\partial I_D}{\partial V_{DS}} = \lambda I_D \tag{3.31}$$

with I_D being either the weak or strong inversion drain current.

An additional transconductance arises if an AC signal is applied to the source with respect to bulk potential. The corresponding *bulk transconductance* g_{mb} is related to the transconductance g_m by

$$g_{mb} = \frac{\partial I_D}{\partial V_{SB}} = g_m \left(n - 1\right) \tag{3.32}$$

3.3.2 Capacitances

The MOSFET exhibits several capacitances of which most are operating point dependent. Table 3.4 lists the gate-source, gate-drain and gate-bulk capacitances together with their operating point dependent values for weak and strong inversion [51, 52].

Capacitance	Weak inversion	Strong Inversion
C_{gs}	$WC_{ov,s}$	$(2/3)\,WLC'_{ox} + WC_{ov,s}$
C_{gd}	$WC_{ov,d}$	$WC_{ov,d}$
C_{gb}	$\dfrac{(WL)C'_{ox}C'_{d}}{C'_{ox}+C'_{d}}$	≈ 0

Table 3.4: MOSFET Capacitances.

The parameters $C_{ov,d}$ and $C_{ov,s}$ characterize capacitances that arise from the gate overlap on the drain and source area, respectively. Figure 3.7 illustrates the simulated values of C_{gs}, C_{gd} and C_{gb} as a function of V_{GS} for a 130 nm technology with L_{min} being the minimal channel length.

In addition, the MOSFET exhibits two depletion region capacitances between bulk and source or drain, C_{sb} and C_{db}, which depend on the applied bulk-source voltage V_{SB} and junction parameters which are hardly accessible for hand calculation.

3.3.3 Figure of Merit

Three figure of merit (FOM) definitions for the MOSFET analog performance are obtained using its small-signal parameters. The first FOM, the transit frequency ω_T, evaluates the high frequency performance of the MOSFET. It is defined as the frequency when the ratio of input current $i_{gs}(j\omega)$ and output current $i_{ds}(j\omega)$ becomes unity [53]. Using (3.28) and (3.30) for the definition of g_m in strong inversion and velocity saturation, the value of C_{gs} in Table 3.4 for strong inversion (neglecting $C_{ov,s}$) and assuming $C_{gs} \approx C'_{ox}$ in velocity saturation [45] gives

$$\left|\frac{i_{ds}(j\omega_T)}{i_{gs}(j\omega_T)}\right| = 1 \Rightarrow \omega_T \approx \frac{g_m}{C_{gs}} = \begin{cases} = \dfrac{3\mu_{eff}V_{DSAT}^{*}}{2L^2}, & \text{for strong inversion} \\ \approx \dfrac{\nu_{sat}}{L}, & \text{for velocity saturation} \end{cases} \tag{3.33}$$

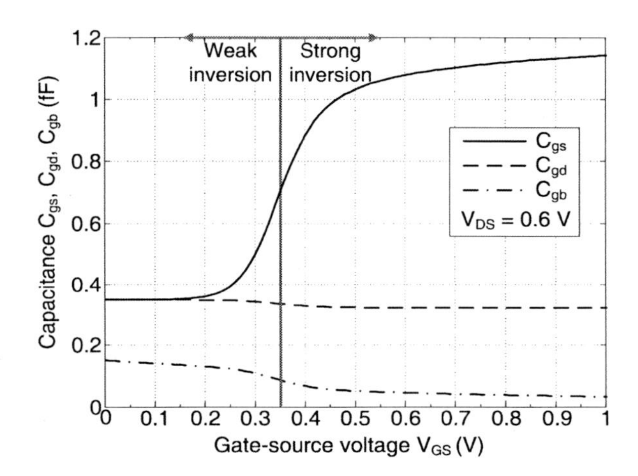

Figure 3.7: Simulated capacitance values of C_{gs}, C_{gd} and C_{gb} versus V_{GS} for a MOSFET in 130 nm technology. A channel width of $W = 10\,L_{min}$ has been used.

Hence, ω_T increases quadratically with technology scaling for long channel devices. For short channel devices velocity saturation becomes pronounced and only a linear increase arises by reducing the devices' channel length L.

A second FOM, important to the design of amplifier, is given by the product of transconductance and output resistance,

$$A_{v0} = g_m r_{ds} = \frac{g_m}{g_{ds}} \tag{3.34}$$

and is called *intrinsic gain* as it represents the maximum DC voltage gain v_{ds}/v_{gs} of a MOSFET common source amplifier (for an infinite load at the drain, e.g. a constant current source).

In general, an increase of the transconductance is expected for future technologies if W/L and $V_{GS} - V_{THn}$ is kept constant in (3.28). Nevertheless, this increase is outperformed by the decrease of output resistance r_{ds} due to downscaling. As a result, A_{v0} decreases for deep-submicron technologies which is demonstrated in Fig. 3.8 showing a comparison of the simulated intrinsic gain for a 350 nm and 130 nm technology. The value of A_{v0} varies between $71.3 - 660$ and $11.8 - 69.5$ for the 350 nm and 130 nm device, respectively. The average ratio of the 350 nm to the 130 nm intrinsic gain lies within $6 \ldots 12.8$.

Finally, a FOM is defined by the ratio of transconduction to drain current gm/I_D. Using (3.28), (3.29) and (3.30) for g_m and the corresponding drain currents given in

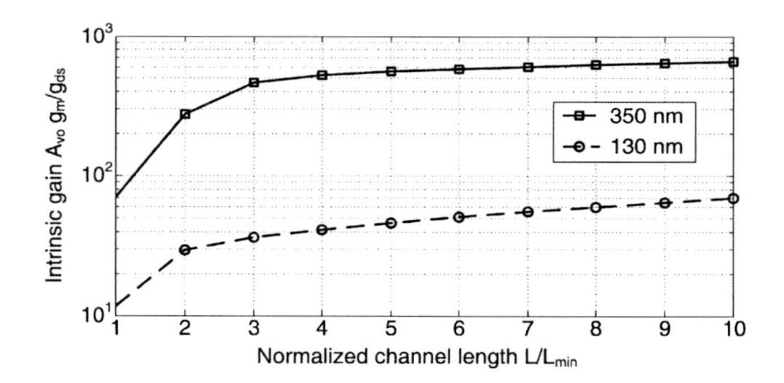

Figure 3.8: Intrinsic gain of a 350 nm and 130 nm NMOS device versus channel length normalized to the minimum gate length L_{min}. The (W/L) ratio has been set to 10 for channel length values, $V_{DS} = V_{DD}/2$ and $V_{GS} = V_{TH} + V_{OV}$ using a V_{OV} of 200 mV.

(3.12), (3.19) and (3.23) results in

$$\frac{g_m}{I_D} = \begin{cases} 1/(V_{GS} - V_{THn}), & \text{for velocity saturation} \\ 2/(V_{GS} - V_{THn}), & \text{for strong inversion} \\ 1/(nU_T), & \text{for weak inversion} \end{cases} \tag{3.35}$$

When plotted with respect to the overdrive voltage it can be shown, that g_m/I_D is nearly technology node independent as indicated by (3.35) [35, 37]. Hence, it can be used as a technology independent measure for the efficiency of the operating point.

3.4 Noise in MOSFETs

3.4.1 Introduction

Electronic *noise* [54, 53, 55] is the spontaneous fluctuation of currents or voltages. In time domain, these fluctuations are expressed using continuous random variables $V_n(t)$ and $I_n(t)$, respectively. The *power spectral density (PSD)* of the noise signals, $v_n^2(f)$ and $i_n^2(f)$, describes the noise power in a 1 Hz bandwidth with units of V^2/Hz and A^2/Hz. Often, the square root of the PSD is given which is called voltage noise or current noise spectral density, denoted as $v_n(f)$ and $i_n(f)$, with units of $V/\sqrt{Hz}$ and $A/\sqrt{Hz}$.

The total noise power within a bandwidth $BW = f_h - f_l$ can be obtained by integration of the PSD, thus

$$v_n^2 = \int_{f_l}^{f_h} v_n^2(f)df \quad \text{and} \quad i_n^2 = \int_{f_l}^{f_h} i_n^2(f)df \tag{3.36}$$

Taking the square root of the total noise power in (3.36) gives the *root mean square* (RMS) value of the noise signals, v_n and i_n, for the bandwidth BW. For a noise signal $V_n(t)$ or $I_n(t)$ that has a Gaussian amplitude distribution with standard deviation σ ($= v_n$ or i_n), the peak-to-peak noise value is given by $6.6\,\sigma$ which ensures that in 99.9% of measurement time no noise values exceeds the calculated peak-to-peak value. For some noise critical applications, σ is being multiplied by a larger value (e.g. $8\,\sigma$).

In order to calculate the response of a linear circuit to two or more uncorrelated noise sources [54], first, the response to each noise source (using its RMS value) is calculated separately with the other voltage noise sources open-circuited and the other current noise sources short-circuited. The resulting RMS values have to be squared, added and the square root of this sum is taken in order to obtain the total RMS noise value of the circuit response.

A fundamental noise source in electronic systems is the thermal noise of resistors. It can be modeled by a noiseless resistor and either a voltage noise source $v_{n,R}(f)$ in series or a current noise source $v_{n,R}(f)$ in parallel as depicted in Fig. 3.9a and 3.9b.

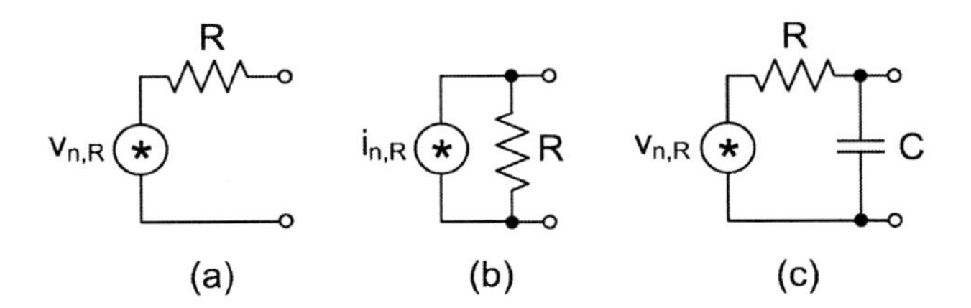

Figure 3.9: Resistor thermal noise model using a voltage noise source (a) and a current noise source (b). (c) Low-pass filter kT/C noise.

The corresponding power spectral densities are given by

$$v_{n,R}^2(f) = 4kTR \tag{3.37}$$

and

$$i_{n,R}^2(f) = \frac{4kT}{R} \tag{3.38}$$

Equations (3.37) and (3.38) are valid at room temperature up to frequencies in the microwave band [54].

The addition of a capacitor in series with a resistor results in a low-pass filter (see Fig. 3.9c) with an integrated power spectral density of kT/C across the capacitor. The integrated PSD is apparently independent of the resistor and is determined by the capacitor size only, hence its name, *kT/C noise*.

The drain current noise of a MOSFET is comprised of separate noise sources. The first one, thermal noise, is due to the channel resistance and has accordingly a flat power spectrum. The second one, flicker or 1/f noise, is due to traps at the $Si - SiO_2$ interfaces. Its PSD is inversely proportional to the frequency, hence the name. At lower frequencies 1/f noise dominates whereas thermal noise contributes mainly to the overall drain current noise. The PSD of the drain current noise is illustrated in Fig. 3.10 with f_k being the transition or *corner frequency* indicating a change of the dominating noise type. Explicit expressions for these noise sources are given hereafter.

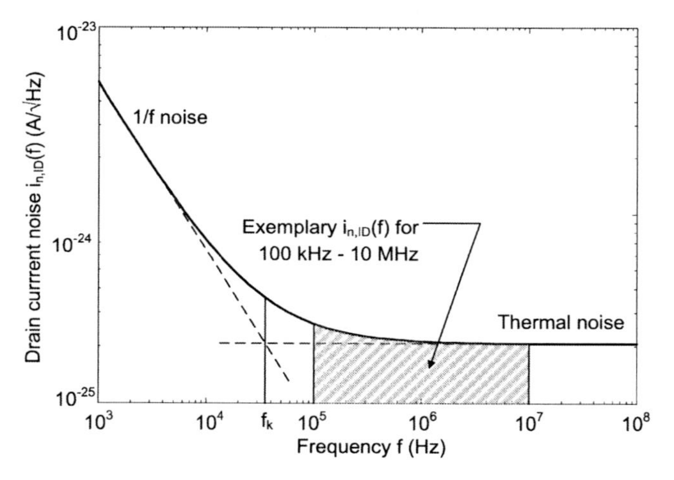

Figure 3.10: MOSFET noise.

3.4.2 Thermal Noise in MOSFETs

A general drain current thermal noise expression, independent of operating region, has the expression [42]

$$i^2_{nth,I_D}(f) = 4kT \, \frac{\mu_{eff} \, |Q_{inv}|}{L^2_{eff}} \tag{3.39}$$

where Q_{inv} is the charge in the inversion layer. There exists yet another general expression relating the thermal noise to the channel resistance at zero drain-source bias g_{d0}, which is [55]

$$i^2_{nth,I_D}(f) = 4kT\,\gamma_n\,g_{do} \tag{3.40}$$

where γ_n is a "noise parameter" or "noise factor". Equation (3.39) is used to simulate thermal noise, whereas (3.40) for performing hand calculations. In the case of strong inversion and for long channel MOSFETs, $g_{do} = g_m$ and $\gamma_n \approx 2/3$, hence

$$i^2_{nth,I_D}(f) = 4kT\,\frac{2}{3}\,g_m \tag{3.41}$$

Dividing (3.41) by g_m^2 results in the input-referred (at V_{GS}) thermal noise which can be written as

$$v^2_{nith,I_D}(f) = 4kT\,\frac{2}{3}\frac{1}{g_m} \tag{3.42}$$

The value of γ_n for short channel devices depends on the technology used and on the MOSFET bias condition. It is slightly higher than 2/3 and lies in the range of $\gamma_n \approx 1$ at low V_{OV} for different short channel technology nodes [56]. Nevertheless, a clear scaling trend cannot be seen from [56] so this topic requires further research.

To derive the drain current thermal noise in weak inversion, (3.29) is substituted in (3.41), giving

$$i^2_{nth,I_D}(f) = \frac{8}{3}\frac{I_D}{n}\,q \approx 2q\,I_D \tag{3.43}$$

and the input-referred value is accordingly

$$v^2_{nith,I_D}(f) = \frac{8}{3}\,kT\,\frac{nU_T}{I_D} \tag{3.44}$$

In addition to the above derivation, two thermal noise sources associated with the gate have to be added [57]. The first one is due to the gate resistance R_G, it is given by

$$v^2_{nith,R_G}(f) = 4kTR_G \tag{3.45}$$

The resistance R_G depends on the layout of the gate and can be therefore minimized by layout techniques. The second one, *induced gate noise*, is due to channel fluctuations inducing a gate current at very high frequencies. Hence, this noise source is usually neglected at low to medium frequencies.

3.4.3 1/f Noise in MOSFETs

The origin of flicker or 1/f noise in MOSFETs has been discussed in numerous publications on the bases of two models. The first one, proposed by McWhorter [58], assumes that fluctuation in the number of free carriers (ΔN) arise from capture and emission of charge carrier at traps (surface states) in the Si-SiO_2 interface. The capture and emission of a single charge carrier results in a random telegraph signal (RTS) having a Lorentzian spectrum. It can be shown that the superposition of Lorentzians results in a 1/f spectrum [59]. The second model, proposed by Hooge [60], relates 1/f noise to fluctuations in device mobility ($\Delta\mu$). The fluctuation of mobility has been attributed to phonon scattering of free charge carriers in the channel [61]. On the one hand the ΔN model has been successfully used to explain 1/f noise in n-channel MOSFETs, whereas on the other hand the $\Delta\mu$ model seems to explain 1/f noise for p-channel devices better [62] (which were buried channel devices at that time).

To combine both models it has been proposed that trapped charge carriers induce correlated mobility fluctuation due to coulomb scattering of free charge carrier in the channel [63]. The resulting model is called *Unified 1/f Noise Model* and results in a drain current 1/f noise of [64]

$$i^2_{nf,I_D}(f) = \frac{kTI_D^2}{\gamma_a f W L}\left(\frac{1}{N_{inv}} + \alpha_s \mu_{eff}\right)^2 N_t \tag{3.46}$$

for low drain-source bias with N_{inv} being the inversion layer charge density, γ_a is a tunneling parameter, α_s is a scattering coefficient and N_t being the effective oxide trap density.

An alternative approach to model 1/f noise is to use the empirical formulation

$$i^2_{nf,I_D}(f) = \frac{KF\, I_D^{AF}}{C'_{ox}\, L^2\, f} \tag{3.47}$$

with KF and $AF \approx 1$ being technology dependent parameters. The input-referred value is then given by

$$v^2_{nif,I_D}(f) = \frac{KF}{2\mu_{0,n}\, C'^2_{ox}\, W L\, f} = \frac{KF'}{C'^2_{ox}\, W L\, f} \tag{3.48}$$

where $KF' = KF/2\mu_{0,n}$.

3.5 Additional Short-Channel Effects

3.5.1 Mismatch

The matching of MOSFETs has already been in focus for long channel technologies. A commonly used analytical model, introduced by Pelgrom, describes the mismatch of a parameter P between two devices [65, 66]:

$$\sigma(\Delta P)^2 = \frac{A_P^2}{W L} + S_P^2 D^2 \tag{3.49}$$

with $\sigma(\Delta P)^2$ being the variance of parameter P, A_P is a proportionality constant associated with the MOSFETs' area $W \times L$ and S_P describes the influence of their spacing D. Two MOSFET parameters are usually considered with respect to mismatch, namely the scaled current factor $\beta = KP(W/L)$ and the threshold voltage V_{TH}. Considering the RMS mismatch values with respect to scaling shows that $\sigma(\Delta\beta)$ stays approximately constant, whereas $\sigma(\Delta V_{THn})$ is showing a proportionality to the gate oxide thickness t_{ox} [65]. The space dependent part of (3.49) can be neglected as wafers become larger with newer technologies thereby reducing variations within a small distance [45]. Hence,

$$\left(\frac{\sigma(\Delta\beta)}{\beta}\right)^2 = \frac{A_\beta^2}{W\,L} \tag{3.50}$$

plays no critical role for modern technologies whereas

$$\sigma(\Delta V_{THn})^2 = \frac{A_{V_{THn}}^2}{W\,L} \tag{3.51}$$

becomes the limiting factor for design in short channel length MOSFETs regarding mismatch.

3.5.2 Gate Leakage Current

Within the sub-100 nm technology nodes, gate leakage current due to tunneling through the gate oxide becomes an important issue for both analog and digital circuits. Using $t_{ox} \approx L_{min}/45$ and a fitting factor α_{leak} [67] and [68, 69],

$$I_{G,Leak} \propto W \left(\frac{V_{GB}}{t_{ox}}\right)^2 e^{\dfrac{-\alpha_{leak}\, t_{ox}}{V_{GB}}} \tag{3.52}$$

and using $L = L_{min}$ shows that any further downscaling increases the gate leakage current significantly. A critical point is reached within the 65 nm node [69], design examples for a regulated cascode in 65 nm show a gain reduction of 30 to 40 dB in comparison to a design in 130 nm [70]

Nevertheless, it is still possible to design analog circuits using appropriate circuit techniques as given by an op amp design in 65 nm [71]. In addition, starting with the 45 nm technology node, the gate oxide material has been changed from SiO_2 to Hafnium based high-k materials. The use of high-k materials shows two benefits. First, a reduction of gate leakage current using a larger t_{ox} and second, a smaller effective gate oxide thickness (EOT) to exploit the benefits of a scaled down technology. In [72], Intel demonstrated that the gate leakage for a 45 nm technology node is comparable to a 130 - 90 nm node using a high-k dielectric.

3.5.3 SCE and RSCE

The threshold voltage of short channel MOSFETs shows a significant gate length dependent deviation from the long channel value V_{T0}. Originally, within the context of *short channel effects (SCE)*, a decrease of V_{TH} was observed for a shorter L or larger V_{DS} due to threshold voltage roll-off and drain-induced barrier lowering (DIBL), respectively. To counteract these, halo/pocket implants are now commonly used thereby adding a reverse SCE (RSCE) in addition to SCE, i.e. V_{TH} increases for shorter L. The influence of both SCE and RSCE on V_{TH} is proportional to [73]

$$V_{TH} \propto V_{T0} - \left(V_{DS}\, e^{-L} - (\zeta - 1)\sqrt{V_{DS}}\, e^{-L/2}\right) \tag{3.53}$$

where $\zeta > 1$ characterizes the pocket implant profile. An exemplary plot of V_{TH} vs. L is shown in Fig. 3.11. For a large V_{DS} both SCE and RSCE are present whereas for a small V_{DS} RSCE dominates which is consistent with (3.53). It has to be noted that the above effects differ significantly between technologies, i.e. for some modern technologies only SCE can be observed.

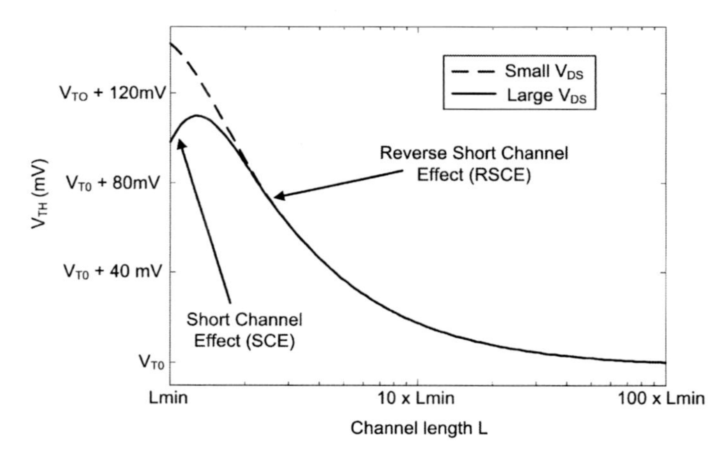

Figure 3.11: Variation of threshold voltage V_{TH} with channel length L due to SCE and RSCE.

Chapter 4

Implementation of Front-End Building Blocks

This chapter describes the design and implementation of the building blocks needed to realize the biomedical signal acquisition front-ends that have been outlined in Section 2.5. In Section 4.1, the implementation of the functional blocks like instrumentation amplifiers or filters is given together with an analysis of their performance specifications and problems associated with non-idealities of these circuits. The outcome of these analysis serves as a criterion to judge the performance of physically implemented blocks. With the operational amplifier playing a central role in the design of the sub-blocks needed, a focus is given on the design of programmable operational amplifier in the following three Sections. The use of programmable operational amplifier enables the design of systems that are flexible with respect to power and noise considerations. In Section 4.2, a general analysis of the programmable operational amplifier will be given. Sections 4.3 and 4.4 describe the design and realization of programmable operational amplifiers using 350 nm and 130 nm, respectively. Finally, application examples for just the 130 nm op amp design are given in Section 4.5, whereas an extensive collection of application examples using one of the 350 nm designs can be found in Chapter 5.

4.1 Functional Block Implementations

4.1.1 3-Op Amp IA

The first instrumentation amplifier (IA) that is described in this Section is the classic 3-op amp IA. Although a 2-op amp IA topology was used for the systems realized in this work it is described here to point out the difference between both topologies and for comparative reasons. Its symmetric architecture is exploited to cancel out op amp non-idealities like frequency dependent open-loop gain. Hence, a relatively high AC CMRR results. This advantage is opposed by the power consumption of three op amps and a larger chip area. Detailed derivations for the gain equations used below are given in Appendix A.1.

Topology

The 3-op amp IA is composed of two stages with the first one being two back-to-back connected non-inverting amplifiers and the second realized as a difference amplifier, see Fig 4.1. The first stage is used to amplify the differential signal whereas a differential- to single-ended conversion is provided by the second stage. A unity common-mode gain is given by the back-to-back connected non-inverting amplifiers which clearly improves the CMRR of the IA.

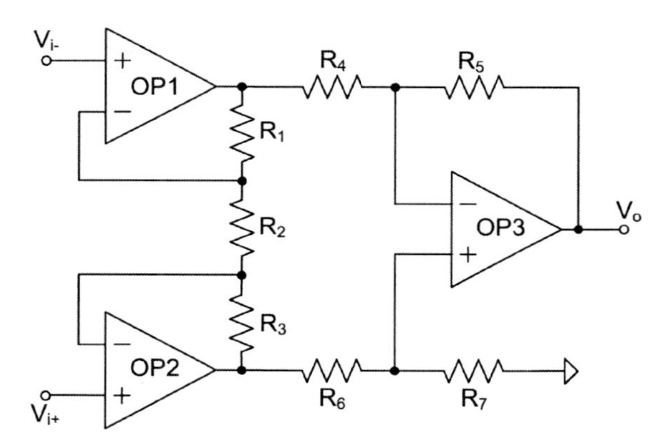

Figure 4.1: 3-op amp IA topology.

Gain

The gain of the difference amplifier is usually set to unity using $R_4 = R_5 = R_6 = R_7$ and the ideal 3-op amp gain becomes

$$A_{3IA} = \frac{V_o}{V_{i+} - V_{i-}} = 1 + 2\frac{R_1}{R_2} \tag{4.1}$$

using $R_3 = R_1$ for the gain setting resistors in the first stage.

The 3-op amp gain error is obtained by analyzing separately the gain error of the first and second stage (see also A.1.1). For the difference amplifier, a relative mismatch in gain setting resistors is given by

$$\frac{R_7}{R_6} = (1 \pm \Delta_{3IA_3})\frac{R_5}{R_4} \tag{4.2}$$

where Δ_{3IA_3} is the relative error in resistance ratio. The gain of the difference amplifier becomes

$$A_3 = \frac{V_o}{V_i} = \frac{1}{2} + \frac{1 \pm \Delta_{3IA_3}}{2 \pm \Delta_{3IA_3}} \tag{4.3}$$

where it has been assumed that $R_4 = R_5$. Likewise,

$$\frac{R_3}{R_2} = (1 \pm \Delta_{3IA_{12}})\frac{R_1}{R_2} \tag{4.4}$$

is used to define the relative error in resistance ratio of the first stage and hence

$$A_{12} = 1 + 2\,\frac{R_1}{R_2} \pm \Delta_{3IA_{12}}\left(\frac{R_1}{R_2}\right). \tag{4.5}$$

where A_{12} denotes the first stage gain. The overall gain is accordingly given by multiplying the gains of the first and second stage:

$$A_{\Delta 3IA} = A_{12} \cdot A_3 = \Bigg(\underbrace{1 + 2\,\frac{R_1}{R_2}}_{A_{3IA}}\underbrace{\pm\Delta_{3IA_{12}}\left(\frac{R_1}{R_2}\right)}_{\Delta A_{12}}\Bigg)\Bigg(1\underbrace{-\frac{1}{2} + \frac{1 \pm \Delta_{3IA_3}}{2 \pm \Delta_{3IA_3}}}_{\Delta A_3}\Bigg) \tag{4.6}$$

Input Common-Mode Range

To begin with, the gain of the first stage will be considered for an analysis of the input-referred common-mode range (ICMR) of the 3-op amp IA. The common-mode gain of the IA equals unity and the differential gain A_{3IA}, hence limiting the output voltages of OP_1 and OP_2 to

$$V_{o1} = A_{3IA}\frac{V_{i,d}}{2} + V_{i,cm} \;\leq\; +V_{DD}/2 \tag{4.7}$$

$$V_{o2} = A_{3IA}\frac{-V_{i,d}}{2} + V_{i,cm} \;\leq\; -V_{DD}/2 \tag{4.8}$$

with differential input voltage $V_{i,d}$, common-mode input voltage $V_{i,cm}$ and using rail-to-rail operational amplifiers. In addition, a signal ground of $V_{DD}/2$ has been assumed. From (4.7) and (4.8), the 3-op amp input common-mode range ICMR_{3IA} is given by

$$\mathrm{ICMR}_{3IA} = -\frac{V_{DD}}{2} + A_{3IA}\frac{V_{i,d}}{2} \ldots + \frac{V_{DD}}{2} - A_{3IA}\frac{V_{i,d}}{2} \tag{4.9}$$

Common-mode Rejection Ratio

The 3-op amp CMRR is determined mainly by the resistor mismatch of the difference amplifier and by the CMRR values of each op amp used in this IA. The frequency dependent open-loop gain, which accounts by a large amount for the CMRR degradation in the 2-op amp IA, is neglected because of the symmetric IA topology. Hence, an overall 3-op amp CMRR is given by [74]

$$\begin{aligned}\frac{1}{\mathrm{CMRR}_{3IA}} = & - \frac{1}{\mathrm{CMRR}_{OP1}} + \frac{1}{\mathrm{CMRR}_{OP2}} \\ & + \frac{1}{A_{3IA}} \frac{1}{\mathrm{CMRR}_{OP3}} + \frac{1}{\mathrm{CMRR}_{\Delta R}} \end{aligned} \tag{4.10}$$

with $\mathrm{CMRR}_{OP1,2,3}$ being the CMRR value of the respective op amps $\mathrm{OP}_{1,2,3}$ and $\mathrm{CMRR}_{\Delta R}$ is the CMRR due to resistor mismatch in the difference amplifier given by

$$\mathrm{CMRR}_{\Delta R} = \frac{\left(R_7\left(R_4+R_5\right)\right)/\left(R_4\left(R_6+R_7\right)\right)+R_4/R_5}{\left(R_7\left(R_4+R_5\right)\right)/\left(R_4\left(R_6+R_7\right)\right)-R_4/R_5} \tag{4.11}$$

With the help of (4.2), Eq. (4.11) can be written in a more compact form, namely

$$\mathrm{CMRR}_{\Delta R} = \frac{4 \pm 3\,\Delta_{3IA_3}}{\pm \Delta_{3IA_3}} \tag{4.12}$$

where it has been assumed, that a difference amplifier with unity gain is used by setting $R_4 = R_5$.

Noise and Power

The 3-op amp noise behavior is obtained by analyzing the difference amplifier and the first stage separately. Because CMOS op amps are used no op amp current noise is considered. The unity gain difference stage with $R_4 = R_5 = R_6 = R_7$ has an input- and output-referred noise PSD of [54]

$$v_{ni3}^2(f) = v_{no3}^2(f) = 4\left(v_{n,R4}^2(f) + v_{ni,OP3}^2(f)\right) \tag{4.13}$$

where $v_{n,R4}^2(f) = 4kTR_4$ and $v_{ni,OP3}^2(f)$ is the input-referred op amp noise PSD.

The first stage output noise is obtained by calculating the contribution of each noise source in Fig. 4.2 to the output of the IA. The resistor R_2 has been divided into two resistors in Fig. 4.2, i.e. $R_{2a} = R_2/2$ and $R_{2b} = R_2/2$, to allow a symmetric analysis of the circuit which yields

$$v_{no1}^2(f) = \left(1+2\frac{R_1}{R_2}\right)^2 v_{ni,OP1}^2(f) + \left(2\frac{R_1}{R_2}\right)^2 v_{n,R2a}^2(f) + v_{n,R1}^2 \tag{4.14}$$

$$v_{no2}^2(f) = \left(1+2\frac{R_1}{R_2}\right)^2 v_{ni,OP2}^2(f) + \left(2\frac{R_1}{R_2}\right)^2 v_{n,R2b}^2(f) + v_{n,R3}^2 \tag{4.15}$$

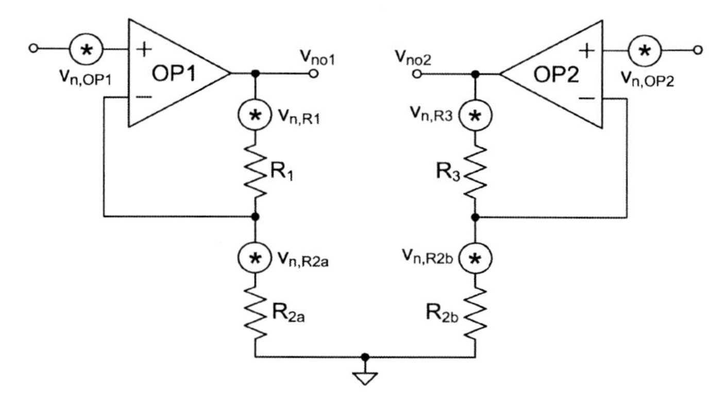

Figure 4.2: First stage of the 3-op amp IA including noise sources.

The first stage output noise $v_{no1}^2(f)$ and $v_{no2}^2(f)$ is fed to the differential amplifier having unity gain and the first stage output noise becomes accordingly

$$\begin{aligned} v_{no12}^2(f) = v_{no1}^2(f) + v_{no2}^2(f) &= \underbrace{\left(1 + 2\frac{R_1}{R_2}\right)^2}_{A_{3IA}^2} \left(v_{ni,OP1}^2(f) + v_{ni,OP2}^2(f)\right) \\ &+ \underbrace{\left(2\frac{R_1}{R_2}\right)^2}_{(A_{3IA}-1)^2} v_{n,R2}^2(f) + 2v_{n,R1}^2(f) \end{aligned} \tag{4.16}$$

The sum of (4.16) and (4.13) is the total output 3-op amp IA noise PSD. Dividing this sum by the gain $A_{3IA} = 1 + 2R_1/R_2$ and assuming all op amps to exhibit the same noise $v_{ni,OP}^2(f)$ gives finally

$$\begin{aligned} v_{ni,IA3}^2(f) = 2\left(1 + \frac{2}{A_{3IA}^2}\right) v_{ni,OP}^2(f) &+ \left(1 - \frac{1}{A_{3IA}}\right)^2 v_{nR2}^2(f) \\ &+ \left(\frac{2}{A_{3IA}^2}\right) \left(v_{n,R1}^2(f) + 2v_{n,R5}^2(f)\right) \end{aligned} \tag{4.17}$$

The static power consumption of the 3-op amp IA is given by

$$P_{3IA,stat} = P_{OP1} + P_{OP2} + P_{OP3} \tag{4.18}$$

and the dynamic power consumption is depending on input signal and the feedback resistor sizes, see also the associated discussion in Section 4.1.2.

4.1.2 2-Op Amp IA

The 2-op amp IA topology uses only 2 op amps which is clearly an advantage with respect to power consumption and chip area. It has therefore been used in the development of the biomedical signal acquisition systems that are described in Chapter 5. Nevertheless, this topology also compromises a strongly frequency dependent CMRR, making it difficult to obtain high CMRR value at the line frequencies of 50/60 Hz. The detailed derivations of the equations for the 2-op amp IA gain, CMRR and noise used in this section can be found in Appendix A.2.

Topology

The 2-op amp IA topology is shown in Fig. 4.3. This architecture can be viewed either as two non-inverting amplifiers connected on top of each other or as two inverting amplifiers connected in series. In the non-inverting amplifier view, V_{i-} is amplified by OP_2 with respect to signal ground and the resulting output voltage is used as the reference voltage for OP_1 which amplifies V_{i+} with respect to this (reference) voltage. In the inverting amplifier view of the IA, signal ground serves as the input voltage to OP_2 and V_{i-} as the reference voltage. The resulting output voltage is fed to the next inverting stage with OP_1 that has now V_{i+} as the reference voltage.

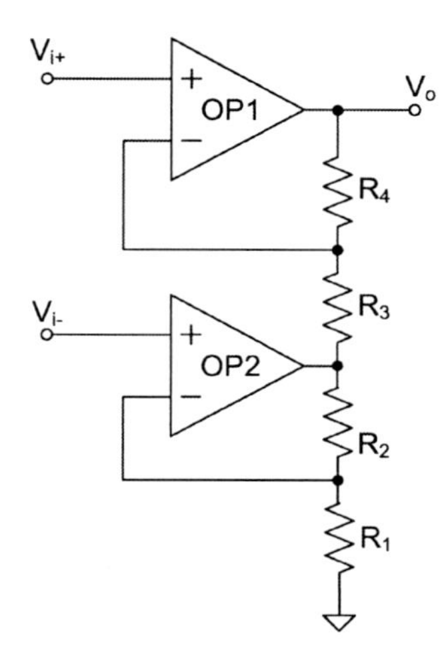

Figure 4.3: 2-op amp IA topology.

Gain

Usually, the feedback resistor values are set to $R_4 = R_1$ and $R_3 = R_2$ in order to obtain a 2-op amp IA gain of

$$A_{2IA} = \frac{V_o}{V_{i+} - V_{i-}} = 1 + \frac{R_1}{R_2} \tag{4.19}$$

Due to mismatch of the resistors, the gain measured deviates from the ideal value. In order to calculate the resulting gain error (see also A.2.1), the relation

$$\frac{R_4}{R_3} = (1 \pm \Delta_{2IA})\frac{R_1}{R_2} \tag{4.20}$$

is used with Δ_{2IA} being the relative mismatch in the ratio of gain setting resistors. Using (4.20) for the derivation of the 2-op amp gain gives

$$A_{\Delta 2IA} = \underbrace{\left(1 + \frac{R_1}{R_2}\right)}_{A_{2IA}} \underbrace{\pm \Delta_{2IA}\left(\frac{1}{2} + \frac{R_1}{R_2}\right)}_{\Delta A_{2IA}} \tag{4.21}$$

with the right hand side determining the gain error due to resistor mismatch.

Input Common-Mode Range

The input common-mode range of the 2-op amp IA is mainly limited by the output voltage of the bottom amplifier OP_2. The gain of the bottom amplifier is given by $A_2 = 1 + R_2/R_1$ and the minimum and maximum output voltage of OP_2 becomes

$$V_{o2} = (V_{i,cm} + V_{i,d}/2) \cdot A_2 = \begin{cases} +V_{DD}/2 \\ -V_{DD}/2 \end{cases} \tag{4.22}$$

assuming a signal ground $V_{sgnd} = V_{DD}/2$ and a rail-to-rail operational amplifier. The input common-mode range is accordingly given by

$$\mathrm{ICMR}_{2IA} = -\frac{V_{DD}}{2A_2} + \frac{V_{i,d}}{2} \dots + \frac{V_{DD}}{2A_2} - \frac{V_{i,d}}{2} \tag{4.23}$$

Common-Mode Rejection Ratio

The CMRR of the 2-op amp IA is determined by its differential-mode gain A_{2IA} given in (4.19) and the common-mode gain $A_{cm} = v_o/v_i$ with v_i being the input common-mode voltage in this case. Fig. 4.4 illustrates the common-mode gain measurement setup, a more detailed analysis can be found in A.2.2.

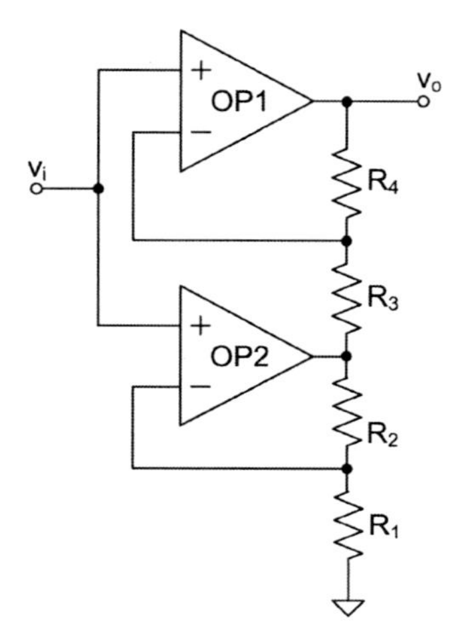

Figure 4.4: 2-op amp IA common-mode gain measurement.

The most significant influence on the 2-op amp IA CMRR is related to the mismatch in resistor ratios R_4/R_1 and R_3/R_2. The common-mode gain influenced by resistor mismatch only, $A_{cm,\Delta R}$, is

$$A_{cm,\Delta R} = \pm\Delta_{2IA} \tag{4.24}$$

where (4.20) has been used for the definition of Δ_{2IA}. The finite 2-op amp CMRR value including the influence of resistor mismatch only (i.e. assuming ideal op amps) is accordingly given by

$$\text{CMRR}_{\Delta R} = \frac{A_{2IA}}{\pm\Delta_{2IA}} \tag{4.25}$$

In addition to this main cause of CMRR limitation due to resistor mismatch, the non-idealities of the op amps have also be taken into consideration. To obtain a quantitative value of this influence on the CMRR of the IA, the common-mode gain is calculated using the assumption of no resistor mismatch which is given by setting $R_4 = R_1$ and $R_3 = R_2$. The op amps are now considered no longer to be ideal but have a finite (frequency dependent) differential gain A_d and a non-zero common-mode gain of A_{cm}. Assuming that $A_{cm}\, R_{1,2} \approx 0$, $A_d\, R_{1,2} \gg R_{1,2}$ and using $A_{cm} = A_d/\text{CMRR}$ results in a 2-op amp IA common-mode gain of

$$\frac{v_o}{v_i} = \underbrace{\frac{(R_1+R_2)^2}{A_{d2} R_1 R_2}}_{A_{cm,\Delta OP}} + \underbrace{\left(1+\frac{R_1}{R_2}\right)\left(\frac{1}{\text{CMRR}_{OP1}} - \frac{1}{\text{CMRR}_{OP2}}\right)}_{A_{cm,\Delta CMRR}} \tag{4.26}$$

with CMRR_{OP1} and CMRR_{OP2} being the CMRR value of OP_1 and OP_2, respectively. Equation (4.26) reveals two sources that contribute to the common-mode gain: First, the term $A_{cm,\Delta OP}$ denotes the common-mode gain resulting from a finite differential gain and second, $A_{cm,\Delta CMRR}$ shows how a difference in the CMRR of the op amps influences the common-mode gain. The contribution of $A_{cm,\Delta OP}$ to the common-mode gain includes the differential open-loop gain of the bottom op amp OP_2. The AC CMRR characteristic can be approximated by modeling the op amp as a simple one pole low-pass with transfer function $A_{d2} = A_{d0}/(1+j(\omega/\omega_p))$ where ω_p denotes the frequency of the pole and A_{d0} is the DC open-loop gain value. The resulting CMRR associated to $A_{cm,\Delta CMRR}$ is given by

$$\text{CMRR}_{\Delta OP} = \frac{A_{2IA}}{A_{cm,\Delta OP}} = \frac{1}{A_2} \cdot \left| \frac{A_{d0}}{1+j(\omega/\omega_p)} \right| \tag{4.27}$$

Fig. 4.5 shows the result of a simulation to illustrate the influence of both frequency and op amp DC open-loop gain for the 2-op amp IA CMRR. It can be seen

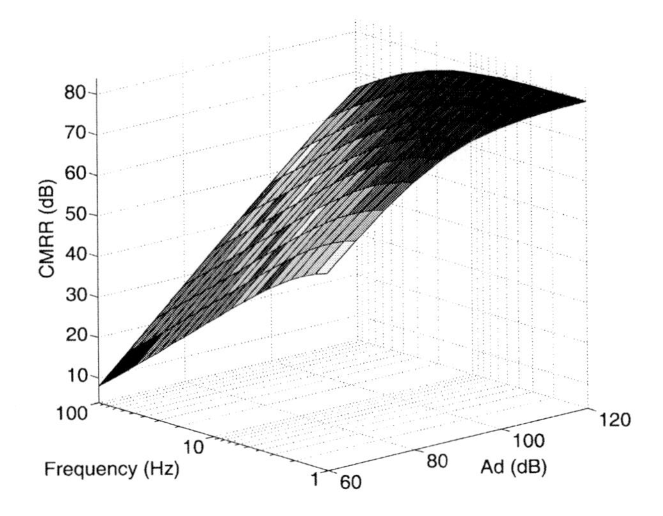

Figure 4.5: Illustration of the 2-op amp IA CMRR dependence on frequency and op amp DC open-loop gain. The parameters used for the simulation are $A_{2IA} = 2$, input common-mode frequencies of $f = 1-100\text{Hz}$, $A_{d0} = 60-120$ dB, $\text{CMRR}_{\Delta R} = 80dB$, $\text{CMRR}_{OP1} = \text{CMRR}_{OP2}$ and the pole frequency is $f_p = 10\text{Hz}$.

clearly how the CMRR degrades as the frequency reaches the pole frequency for high A_d values. An additional CMRR degradation is observed for lower A_d values.

Finally, considering the CMRR value associated with $A_{cm,\Delta CMRR}$ only gives

$$\mathrm{CMRR}_{\Delta CMRR} = \frac{A_{2IA}}{A_{cm,\Delta CMRR}} = \left(\frac{1}{\mathrm{CMRR}_{OP1}} - \frac{1}{\mathrm{CMRR}_{OP2}}\right)^{-1} \tag{4.28}$$

This value depends clearly on the mismatch between the CMRR values of both op amps. An interesting point to note is that $\mathrm{CMRR}_{\Delta CMRR}$ can be either larger or smaller than 0. Any mismatch in op amp CMRR can therefore either increase or decrease the overall CMRR performance of the 2-op amp IA.

The overall 2-op amp IA CMRR is found by superposing the reciprocals of the partial common-mode rejection ratios [39]. Thus, using (4.25), (4.27) and (4.28),

$$\frac{1}{\mathrm{CMRR}_{2IA}} = \frac{A_{cm,\Delta OP}}{A_{2IA}} + \frac{A_{cm,\Delta CMRR}}{A_{2IA}} + \frac{A_{cm,\Delta R}}{A_{2IA}} \tag{4.29}$$

$$= \frac{1}{\mathrm{CMRR}_{\Delta OP}} + \frac{1}{\mathrm{CMRR}_{\Delta CMRR}} + \frac{1}{\mathrm{CMRR}_{\Delta R}} \tag{4.30}$$

In order to demonstrate the influence of $\mathrm{CMRR}_{\Delta OP}$ and $\mathrm{CMRR}_{\Delta R}$ on the overall 2-op amp IA common-mode rejection ratio in (4.30), a simulation was performed by sweeping the frequency and setting the CMRR due to resistor mismatch to a constant value of 80 dB. Additionally, it was assumed that both op amps exhibit the same CMRR, i.e. $\mathrm{CMRR}_{\Delta CMRR} = 0$. The remaining values were set to $A_{d0} = 80$ dB, $f_p = 100$ Hz and $A_{2IA} = 4$. The simulation result in Fig. 4.6 shows how the overall CMRR decreases for higher frequencies as the op amp open-loop gain decreases, whereas it is dominated by $\mathrm{CMRR}_{\Delta R}$ for low frequencies.

Noise and Power

The 2-op amp IA noise sources are shown in Fig. 4.7. The noise contribution from the feedback resistors is given by $v_{n,R}^2(f) = 4kTR$ and the op amp noise is represented by $v_{ni,OP}^2(f)$.

The 2-op amp IA output noise PSD becomes (more details are given in A.2.3)

$$v_{no,2IA}^2(f) = 2\left[\underbrace{\left(1+\frac{R_1}{R_2}\right)}_{A_{2IA}} v_{n,R1}^2(f) + \underbrace{\left(1+\frac{R_1}{R_2}\right)^2}_{A_{2IA}^2} v_{ni,OP}^2(f)\right] \tag{4.31}$$

using $R_4 = R_1$ and $R_3 = R_2$. Dividing this result by the squared gain of the IA gives the equivalent input-referred value:

$$v_{ni,2IA}^2(f) = 2\left(\frac{v_{n,R1}^2(f)}{A_{2IA}} + v_{ni,OP}^2(f)\right) \tag{4.32}$$

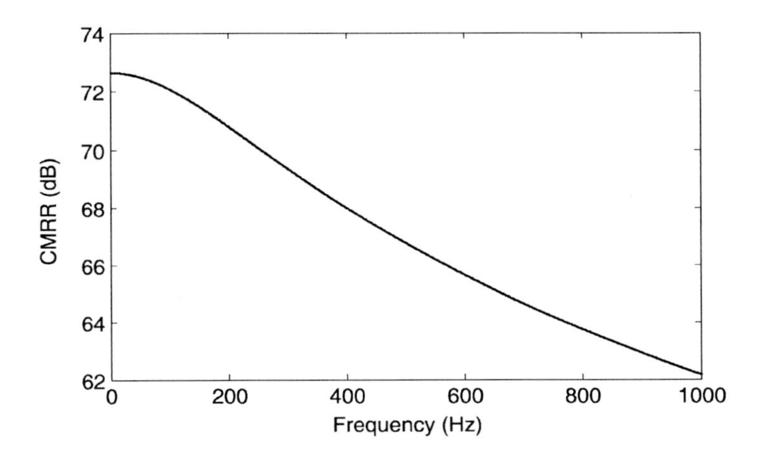

Figure 4.6: CMRR of the 2-op amp IA versus frequency for a constant resistor mismatch. An op amp macro-model has been used for the simulation.

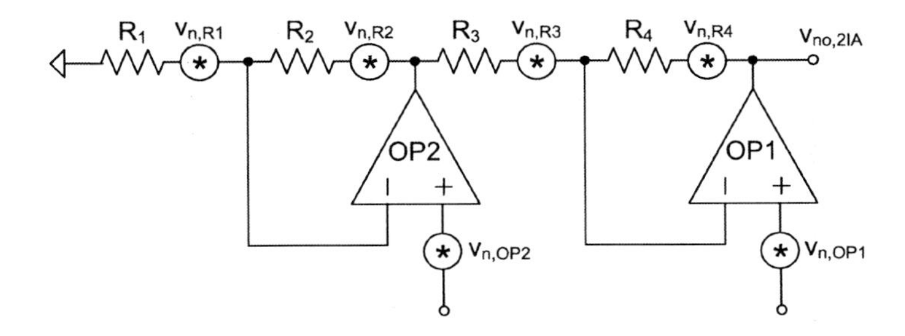

Figure 4.7: 2-op amp IA including noise sources.

The static 2-op amp IA power consumption is given by

$$P_{2IA,stat} = P_{OP1} + P_{OP2} \tag{4.33}$$

and the dynamic power consumption is determined by the size of the feedback resistors and the input signal which is, however, not exactly known before. An optimal value for the size of the feedback resistors is given by the largest resistor size which keeps the thermal noise of the resistors still significantly smaller than the overall 2-op amp IA noise. The same arguments also holds for the 3-op amp IA, PGA and the low-pass filter. Fig. 4.8 illustrates this noise-power trade-off for an 2-op amp IA by showing the simulated input-referred thermal noise and the power dissipation for feedback resistors $R_1 = R_2$ being swept from 100 Ω to 1 MΩ. The input signal V_i was set to a constant value of 0.5 V. It can be seen that an optimal feedback resistor size is given in the range of kilohms where the thermal noise almost reaches its lowest value.

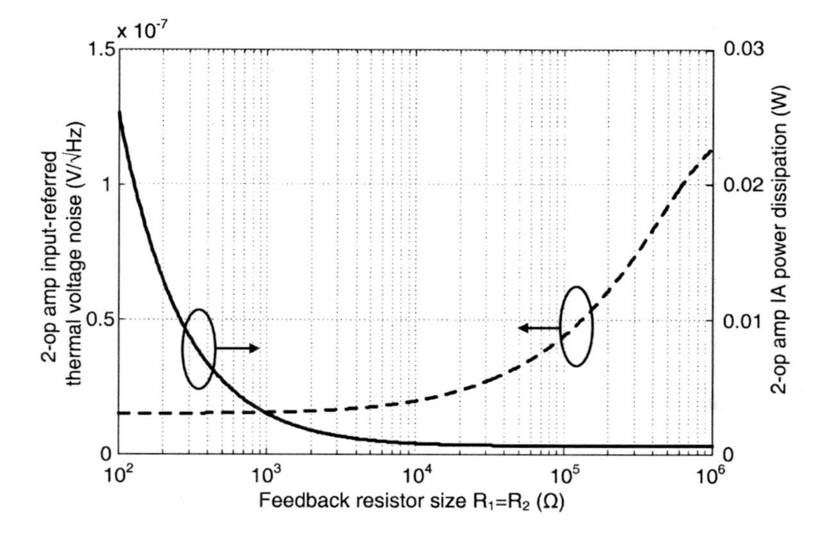

Figure 4.8: 2-op amp IA noise vs. power trade-off for changing feedback resistor sizes.

4.1.3 PGA(Postamp)

Topology and Gain

A common realization of the programmable gain amplifier (PGA) is composed of an op amp in an non-inverting configuration. This type of amplifier is usually acting as a second gain stage in addition to the IA situated at the very front of the signal acquisition channel, hence it is also referred to as the postamplifier. Within the scope of this work, the term PGA will be used mainly.

The gain setting can be realized by changing the ratio of the resistors $R_1 = R_{1a} + R_{1b}$ and $R_2 = R_{2a} + R_{2b}$ as shown in Fig. 4.9. and the PGA gain is given by

$$A_{PGA} = 1 + \frac{R_{1a} + R_{1b}}{R_{2a} + R_{2b}} \tag{4.34}$$

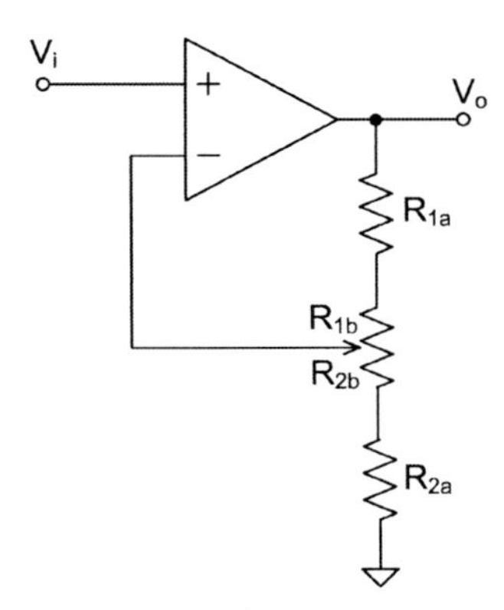

Figure 4.9: PGA topology.

The resistor ratio change can be realized by a resistor chain having several taps that are connected to the inverting input of the op amp using a switch. Turning one switch on and all others off changes the gain accordingly. The switches are implemented as CMOS transmission gates that can be turned on and off using a digital control line.

The gain error of the PGA is directly related to the mismatch in resistor ratio R_1/R_2.

Noise and Power

The PGA noise PSD is given by the noise calculation of an op amp in an non-inverting amplifier configuration having an output-referred value of [54]

$$v_{no,PGA}^2(f) = \underbrace{\left(1+\frac{R_1}{R_2}\right)^2}_{A_{PGA}^2} v_{ni,OP}^2(f) + \underbrace{\left(\frac{R_1}{R_2}\right)^2}_{(A_{PGA}-1)^2} v_{n,R2}^2(f) + v_{n,R1}^2(f) \tag{4.35}$$

with $v_{ni,OP}^2(f)$ and $v_{n,R}^2(f)$ being defined as in the 3/2-op amp IA noise calculations. Dividing (4.35) by the squared gain A_{PGA}^2 gives an input-referred noise PSD of

$$v_{ni,PGA}^2(f) = v_{ni,OP}^2(f) + \left(1-\frac{1}{A_{PGA}}\right)^2 v_{n,R2}^2(f) + \frac{v_{n,R1}^2(f)}{A_{PGA}^2} \tag{4.36}$$

The static power consumption of the PGA $P_{PGAOP,stat}$ is determined by the static op amp power dissipation. For the dynamic power dissipation refer to the discussion in Section 4.1.2.

4.1.4 Low-Pass Filter

There exists a variety of architectures to implement a low-pass filter like op amp based active filters, gm-c filters or switched capacitor filters. This multitude of realizations hinders to set up general considerations regarding the circuit performance. Nevertheless, one particular realization using an op amp, the *unity gain Sallen-Key low-pass filter*, is described here because of its use in the realized systems of Sections 5.2 and 5.3.

Topology and Transfer Function

The unity gain Sallen-Key low-pass filter shown in Fig. 4.10 [75] realizes a two pole system whose generalized transfer function is given by

$$A_{LP,gen}(s) = \frac{1}{\left(\frac{s}{\omega_c}\right)^2 + \frac{1}{Q}\left(\frac{s}{\omega_c}\right) + 1} \tag{4.37}$$

with $\omega_c = 2\pi f_c$ being the cut-off frequency and Q is the quality factor that determines the peaking of the system. The value of Q sets also the filter type, e.g for $Q = 0.707$ we obtain a Butterworth type, for $Q = 1.305$ a 3-dB Tschebyscheff type and critical damping for $Q = 0.5$.

The transfer function of the unity gain Sallen-Key low-pass depicted in Fig. 4.10 can be calculated using $s = j\omega$ as

$$A_{SKLP}(j\omega) = \frac{1}{(j\omega)^2 R_1 R_2 C_1 C_2 + j\omega C_1(R_1+R_2)+1} \tag{4.38}$$

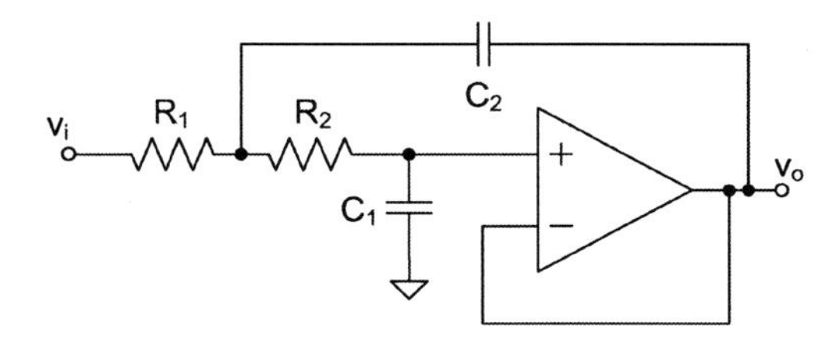

Figure 4.10: Sallen-Key low-pass topology.

To simplify the filter design, the resistors and capacitors are set as ratios, i.e. $m = R_2/R_1$ and $n = C_2/C_1$ which results in a cut-off frequency of

$$\omega_c = \frac{1}{2\pi R_1 C_1 \sqrt{mn}} \tag{4.39}$$

and a quality factor that is given by

$$Q = \frac{\sqrt{mn}}{m+1} \tag{4.40}$$

Noise and Power

The output noise of the Sallen-Key low-pass filter is comprised of three noise sources as shown in Fig. 4.11a, i.e. the thermal noise PSD of the resistors $v_{n,R1}^2(f)$ and $v_{n,R2}^2(f)$ and the op amp noise PSD $v_{ni,OP}^2(f)$. The op amp output noise contributes directly to the overall output noise $v_{no,SKLP}^2(f)$, whereas $v_{n,R1}^2(f)$ and $v_{n,R2}^2(f)$ are filtered by the action of the low-pass. The transfer function of only$v_{n,R1}(f)$ to the output is given by (4.38), i.e. $v_{no,R1}(j2\pi f) = A_{SKLP}(j2\pi f) \cdot v_{n,R1}(f)$ using $\omega = 2\pi f$.

In order to calculate the transfer function of only $v_{n,R2}(f)$ to the output the equivalent circuit shown in Fig. 4.11b is used. Equating the currents i_{C2}, i_{R1} and $i_{R_2C_1}$ at the node v_x we obtain

$$\begin{aligned} i_{C2} &= i_{R1} + i_{R2C1} \\ &\Leftrightarrow \\ \frac{v_{no,SKLP}(f) - v_x}{1/j2\pi f C_1} &= \frac{v_x + v_{n,R2}(f)}{R_2 + 1/j2\pi f C_2} + \frac{v_x}{R_1} \end{aligned} \tag{4.41}$$

and

$$\frac{v_x + v_{n,R2}(f)}{v_{no,SKLP}(f)} = \frac{R_2 + 1/j2\pi f C_1}{1/j2\pi f C_1} \tag{4.42}$$

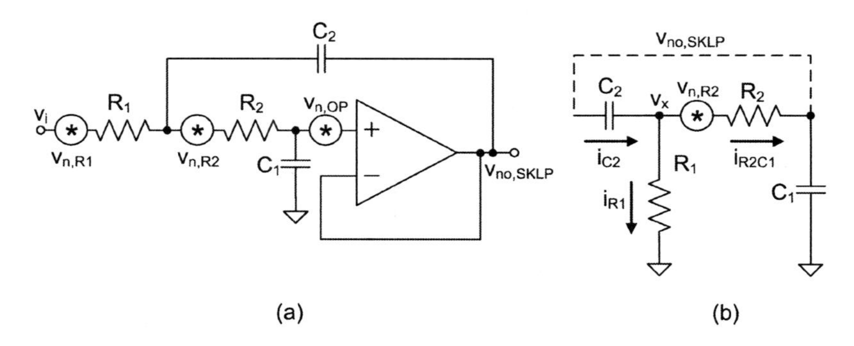

Figure 4.11: Sallen-Key noise analysis. (a) Full circuit including all noise sources. (b) Equivalent circuit to determine the output noise of R_2.

Inserting (4.42) into (4.41) and solving for $v_{no,SKLP}/v_{n,R2}$ gives a transfer function of

$$v_{n,R2}(j2\pi f) = \frac{1 + j2\pi f R_1 C_2}{(j2\pi f)^2 R_1 R_2 C_1 C_2 + j2\pi f C_1(R_1 + R_2) + 1} \cdot v_{n,R2}(f) \tag{4.43}$$

The total integrated output voltage noise power is accordingly given by

$$v_{no,SKLP}^2 = \int_0^\infty |A_{nR1}(j2\pi f)|^2 + |A_{nR2}(j2\pi f)|^2 + |v_{ni,OP}(j2\pi f)|^2 \, df \tag{4.44}$$

The static power consumption of the unity gain Sallen-Key low-pass is given by the static power dissipation of the op amp used. A rough estimate of the dynamic power dissipation of the low-pass is obtained by considering input signals with frequencies lying in the pass band only (i.e. $f < f_c$). Then, like for the static case, the op amp determines the overall power dissipation.

4.2 Programmable Op Amp Design

The op amps needed to realize the functional blocks of Section 4.1 differ in their specification requirements. In the IA they have high demands regarding their (input-referred) noise as well as maximum power dissipation. In contrast to this only low bandwidth requirements are given by the nature of the biomedical signals to be acquired. Furthermore, the ratio between input-referred noise and maximum power dissipation is application dependent, i.e. for long-term mobile ECG systems power requirements dominate whereas for extremely sensitive EP recordings low-noise amplification becomes essential. The op amp in the PGA and low-pass have rather average noise and power constraints, they require however a high slew rate and larger output currents. In addition, all op amp should exhibit a rail-to-rail input and output behavior to maximize the input common-mode range of the system.

A versatile solution to these different requirements is given by the use of a programmable op amp. This type of op amp has the ability to work within a programming range with a low-noise mode at one end of the programming range and a low-power mode at the other. The programming is realized by changing the bias currents of the op amp which influence also other specifications like bandwidth or slew rate as shown in Table 4.1 [76].

	$I_{bias} \Uparrow$
Slew rate	$\Uparrow$
Bandwidth	$\Uparrow$
Input voltage noise	$\Downarrow$
Power dissipation	$\Uparrow$
Open-loop gain	$\Updownarrow$

Table 4.1: Influence of the external bias current on op amp parameters [76].

4.2.1 General Approach and Op Amp Topology

The programmable op amps described in this work are based on a rail-to-rail topology presented by Hogervorst et al. in 1994 [77] with a detailed description of this topology given in [78]. A CMOS 350 nm programmable op amp design using this topology has been described by Bronskowski et al. [30]. Figure 4.12 shows the basic topology of the op amp which is comprised of a rail-to-rail input stage (M1-M4) and folded-cascode stage including class-AB biasing (M5-M20) for the output stage (M21-M22). Two different frequency compensations are depicted in Fig. 4.12 using capacitors C_{c1} and C_{c2}. The first one (continuous lines) is a simple Miller compensation scheme and the second one (dotted lines) shows the cascoded Miller compensation. The NMOS and PMOS input pairs are each biased by an external bias current I_{diff} and the cascode and output stage is biased by two current sources I_{ab}.

4.2.2 Rail-to-Rail Folded-Cascode Input Stage

The rail-to-rail folded cascode input stage is comprised of the differential pairs M1-M4, cascode transistors M9-M12 and two current mirrors M5-M8 that were incorporated into the cascode stage. The bias current of the cascode stage is generated by

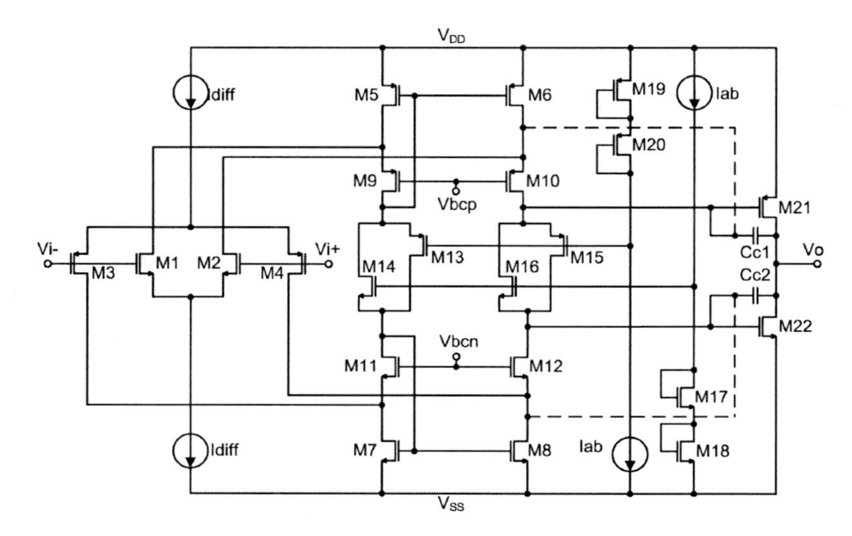

Figure 4.12: Rail-to-rail op amp topology with Miller compensation (the dotted lines indicate the version with cascoded Miller compensation).

a floating current source which is formed by M13/M14 and associated bias voltage generation using M17-M20.

The minimum common-mode input voltage of the NMOS pair is given by its gate-source voltage V_{GSn} and the overdrive voltage V_{OV} needed to keep the I_{diff} current source (which is implemented by at least one transistor) in saturation, hence

$$\min \{V_{i,cm}\}_n = V_{GSn} + V_{OV} + V_{SS} = V_{THn} + 2V_{OV} + V_{SS} \tag{4.45}$$

The maximum NMOS pair common-mode input voltage is limited by the overdrive voltages V_{OV} between drain and source of the NMOS pair and M5/M6, therefore

$$\max \{V_{i,cm}\}_n = V_{DD} - 2V_{OV} + V_{GSn} = V_{DD} + V_{THn} - V_{OV} \tag{4.46}$$

An analysis of the minimum and maximum input common-mode voltage for the PMOS pair yields likewise

$$\min \{V_{i,cm}\}_p = -V_{SGp} + V_{OV} + V_{SS} = -V_{THp} + V_{OV} + V_{SS} \tag{4.47}$$

and

$$\max \{V_{i,cm}\}_p = V_{DD} - V_{SGp} - V_{OV} = V_{DD} - V_{THp} - 2V_{OV} \tag{4.48}$$

The common-mode input voltage of this rail-to-rail input stage can accordingly exceed the supply rails, it has nevertheless one major drawback which is given by the fact that its input transconductance g_m drops by a factor of two when only one input pair operates. This is the case for input common-mode voltages larger than $\max\{V_{i,cm}\}_p$ and lower than $\min\{V_{i,cm}\}_n$, i.e. the NMOS pair can reach the positive rail whereas the PMOS pair does not and vice versa. The large variation in g_m gives rise to distortion and prevents an optimal frequency compensation with respect to power dissipation. To circumvent this problem an additional *constant-g_m* circuitry has to be added to the input stage which is detailed for the 130 nm op amp in Section 4.4.2.

The function of the rail-to-rail folded cascode input stage can be described as follows. The AC currents generated by the input transistors M2 and M4 are directly fed to the drains of M10 and M12, respectively. For the AC currents generated by M1 and M3 two current mirrors incorporated in the cascode copy these currents from the left hand side of the cascode stage to the right hand side and, finally, also to the drains of M10 and M12. The gain of the input stage is determined by its g_m times the small-signal resistance that is seen at the drain nodes of M10 and M12.

The current of the floating current source $I_{FCS} = I_{D13} + I_{D14}$ is set by two translinear loops, namely $V_{gs17} + V_{gs18} = V_{gs14} + V_{gs7}$ and $V_{gs19} + V_{gs20} = V_{gs13} + V_{gs5}$. Assuming all MOSFETs being in saturation, each transistor type (NMOS and PMOS, respectively) having a KP and threshold voltage of about the same size and using (3.12) gives

$$\sqrt{\frac{I_{ab}}{(W/L)_{17}}} + \sqrt{\frac{I_{ab}}{(W/L)_{18}}} = \sqrt{\frac{I_{FCS}/2}{(W/L)_{14}}} + \sqrt{\frac{I_{FCS}}{(W/L)_{7}}} \tag{4.49}$$

$$\sqrt{\frac{I_{ab}}{(W/L)_{19}}} + \sqrt{\frac{I_{ab}}{(W/L)_{20}}} = \sqrt{\frac{I_{FCS}/2}{(W/L)_{13}}} + \sqrt{\frac{I_{FCS}}{(W/L)_{5}}} \tag{4.50}$$

where it has been assumed that $I_{D13} = I_{D14} = I_{FCS}/2$. Rearranging (4.49) and (4.50) with respect to I_{FCS} gives

$$I_{FCS} = \gamma_{OP} I_{ab} \tag{4.51}$$

where

$$\gamma_{OP} = \left(\frac{1/(W/L)_{17} + 1/(W/L)_{18}}{1/2(W/L)_{14} + 1/(W/L)_{7}}\right)^2 = \left(\frac{1/(W/L)_{19} + 1/(W/L)_{20}}{1/2(W/L)_{13} + 1/(W/L)_{5}}\right)^2 \tag{4.52}$$

A simplification to this expression can be made by setting the (W/L) ratios equal between corresponding PMOS and NMOS transistors, e.g. $(W/L)_{22}/(W/L)_{21} = (W/L)_{17}/(W/L)_{2}$. Care has to be taken for designs in a technology where the n-channel and p-channel MOSFETs do not exhibit the same threshold voltages, respectively. In these cases only gross approximations are obtained using the above equations.

4.2.3 Output Stage

The MOSFETs M21 and M22 of the output stage have to be connected in a common-source configuration in order to obtain a rail-to-rail output swing. Additionally, none of the output transistors should turn off completely at full output swing to prevent any distortions. This task is accomplished by the feedforward class-AB control consisting of M15-M20. The translinear loops $V_{gs17} + V_{gs18} = V_{gs16} + V_{gs22}$ and $V_{gs19} + V_{gs20} = V_{gs15} + V_{gs21}$ control the output current of the op amp. Using the same assumptions like in (4.49) and (4.50) we get

$$\sqrt{\frac{I_{ab}}{(W/L)_{17}}} + \sqrt{\frac{I_{ab}}{(W/L)_{18}}} = \sqrt{\frac{I_{D16}}{(W/L)_{16}}} + \sqrt{\frac{I_{D22}}{(W/L)_{22}}} \tag{4.53}$$

$$\sqrt{\frac{I_{ab}}{(W/L)_{19}}} + \sqrt{\frac{I_{ab}}{(W/L)_{20}}} = \sqrt{\frac{I_{D15}}{(W/L)_{15}}} + \sqrt{\frac{I_{D21}}{(W/L)_{21}}} \tag{4.54}$$

The currents in M15 and M16 are fixed by the condition that

$$I_{D15} + I_{D16} = I_{D12} = I_{D10} = \gamma_{OP} I_{ab} \tag{4.55}$$

where it has been assumed that the currents I_{D12} and I_{D10}, respectively, equal the current generated by the floating current source I_{FCS} which is given by (4.51) and (4.52). The quiescent current of the output stage I_q is obtained by setting $I_{D15} = I_{D16} = \gamma_{OP} I_{ab}/2$ and solving (4.53) and (4.54) for I_{D21} and I_{D22}:

$$I_q = \delta_{OP} I_{ab} \tag{4.56}$$

where

$$\begin{aligned} \delta_{OP} &= (W/L)_{22} \left(\frac{1}{(W/L)_{17}} + \frac{1}{(W/L)_{18}} - \frac{\gamma_{OP}}{2(W/L)_{16}} \right)^2 \\ &= (W/L)_{21} \left(\frac{1}{(W/L)_{19}} + \frac{1}{(W/L)_{20}} - \frac{\gamma_{OP}}{2(W/L)_{15}} \right)^2 \end{aligned} \tag{4.57}$$

The minimum output stage current of this class-AB circuit is set by the condition (4.55) and assuming that one of the transistors M15 or M16 carries all the current $\gamma_{OP} I_{ab}$ while the other is shut off. As a result, the factor two in front of $(W/L)_{15}$

or $(W/L)_{16}$ in (4.57) has to be dropped in the calculation of (4.56), which is in this case the minimum output stage current.

Again, the calculations can be greatly simplified by taking equal (W/L) ratios between corresponding transistors and care has to be taken when using a technology with different threshold voltages for NMOS and PMOS, respectively.

Finally, it should be noted that although named rail-to-rail output stage, it is not possible to fully reach the supply rails because of M21 and M22 going from saturation into linear region for very small drain-source voltages. An approximation of the output swing is given by [78]:

$$V_{o,min} = -\frac{V_{DD}}{2}\left(1 - \frac{1}{(KP_n/2)(W/L)_{22}V_{OV}R_L + 1}\right) \tag{4.58}$$

and

$$V_{o,max} = \frac{V_{DD}}{2}\left(1 - \frac{1}{(KP_p/2)(W/L)_{21}V_{OV}R_L + 1}\right) \tag{4.59}$$

where R_L denotes a load resistor connected to the output.

4.2.4 Small-Signal Equivalent Circuit

The small-signal equivalent circuit of the Miller compensated op amp shown in Fig. 4.13 does not basically differ from a generic 2-stage Miller compensated amplifier when incorporating the small-signal characteristics of the cascode transistors M9 – M12 into the output resistance r_A.

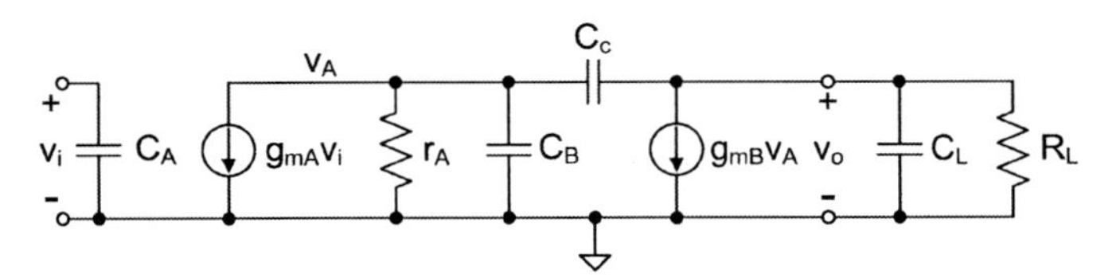

Figure 4.13: Small-signal equivalent circuit.

The equivalent circuit consists of the op amp input stage capacitance

$$C_A = C_{gs1} + C_{gs2} + C_{gs3} + C_{gs4} \tag{4.60}$$

and transconductance

$$g_{mA} = g_{m1} + g_{m2} + g_{m3} + g_{m4} = 4\,g_{m1} \tag{4.61}$$

where it has been assumed that $v_i/2$ is applied at each each input transistor and $g_{m1} = g_{m2...4}$, respectively.

The output resistance r_A is the parallel connection of the resistances that are seen from the drains of M10 and M12, respectively, which implies that M15 and M16 of the class-AB control circuit are treated like a DC voltage source which is replaced accordingly by a short circuit in the equivalent circuit. The resistances seen from the drains of the cascode transistors M12 and M10 are approximately the output resistance of M6 and M8 multiplied by the intrinsic gain of M10 and M12 [39], hence

$$r_A \approx ((g_{m12} \cdot r_{ds12})\, r_{ds8}) \,||\, ((g_{m10} \cdot r_{ds10})\, r_{ds6}) \tag{4.62}$$

$$= ((g_{m12} \cdot r_{ds12})\, r_{ds8})\, /2 \tag{4.63}$$

assuming equal transconductances and output resistances between corresponding NMOS and PMOS devices. The second stage is modeled by its transconductance

$$g_{mB} = g_{m21} + g_{m22} = 2\, g_{m22} \tag{4.64}$$

with $g_{m22} = g_{m21}$ and its input capacitance which is approximated by

$$C_B = C_{gs21} + C_{gs22} + C_{db22} + C_{db10} + C_{db12} \tag{4.65}$$

$$C_{db15} + C_{db16}$$

$$\approx 3\,(C_{gs21} + C_{gs22}) \tag{4.66}$$

The drain-bulk capacitances in (4.66) were replaced by the approximation that these equal the gate-source capacitances in order to obtain a simply expression [45].

The small-signal equivalent circuit of Fig. 4.13 with no Miller capacitor C_c is characterized by two poles. Introducing the capacitor C_c $(= C_{c1} + C_{c2})$ moves these poles apart with the dominant pole ω_{p1} and the non-dominant pole ω_{p2} being moved to lower and higher frequencies, respectively. Furthermore, a zero is also added.

An analysis of the small-signal circuit of Fig. 4.13 [53, 45, 78] yields that the dominant and non-dominant poles are located at the frequencies

$$\omega_{p1} = -\frac{1}{g_{mB} R_L r_A C_c} \tag{4.67}$$

and

$$\omega_{p2} \approx -\frac{g_{mB}}{C_L \left(1 + \dfrac{C_A}{C_c}\right)} \tag{4.68}$$

and the zero is located at

$$\omega_z = \frac{g_{mB}}{C_c} \tag{4.69}$$

In order to derive the DC open-loop gain, the DC gain of the input stage $g_{mA} r_A$ has to be multiplied by the DC gain of the output stage $g_{mB} R_L$, hence

$$A_{d0} = g_{mA} r_A g_{mB} R_L \tag{4.70}$$

The transfer function of the small-signal equivalent circuit in Fig. 4.13 is given by

$$\frac{v_o(s)}{v_i(s)} = A_{d0} \frac{\left(1 - \frac{s}{\omega_z}\right)}{\left(1 + \frac{s}{\omega_{p2}}\right)\left(1 + \frac{s}{\omega_{p1}}\right)} \tag{4.71}$$

with the unity gain frequency located at

$$\omega_u = \frac{g_{mA}}{C_c} \tag{4.72}$$

A small-signal equivalent circuit and associated (stability) analysis of the cascoded Miller compensation is presented in [30] and is therefore not given here. In sum, the cascoded Miller compensating extends the bandwidth of the op amp by a factor of C_c/C_{gs1} whereas a drawback of this compensation scheme is that gain peaking can occur at frequencies higher than the unity gain frequency, which endangers the stability of the amplifier.

4.2.5 Stability

The stability of the Miller compensated op amp is analyzed using the equivalent circuit in Fig. 4.13. The dominant pole ω_{p1} of the transfer function (4.71) introduces a 90° phase shift at the unity gain frequency ω_u. An additional phase shift has to be added by the non-dominant pole ω_{p2} and zero ω_z which are assumed to lie at frequencies greater than ω_u. The op amp phase margin PM is therefore determined by this additional phase shift [79]:

$$PM = 90° - \arctan\left(\frac{\omega_u}{|\omega_{p2}|}\right) - \arctan\left(\frac{\omega_u}{\omega_z}\right) \tag{4.73}$$

To avoid peaking of the op amp when operating as a voltage follower, i.e. having a gain of one, its closed loop transfer function should have a phase margin of $PM_{stab} \approx 60° - 70°$ which can be derived from (4.37) and using a quality factor of $Q \approx 0.7$ [45]. Using (4.68), (4.69) and (4.72) a condition for the stability of the Miller compensated op amp is hence given by

$$\arctan\left(\frac{g_{mA}}{g_{mB}}\frac{C_L}{C_c}(1 + C_A/C_c)\right) + \arctan\left(\frac{g_{mA}}{g_{mB}}\right) = 90° - PM_{stab} \tag{4.74}$$

4.2.6 Noise

To begin with the noise analysis of the presented op amp topology it is considered that the op amp is comprised of two cascaded amplifier stages. The second stage output noise is hence divided by the gain of both stages to refer the noise to the op amp input whereas the first stage noise is divided by the gain of the first stage only. It can be concluded that neglecting the second stage noise does not introduce a large error in the noise analysis as it becomes attenuated by several orders of magnitude in comparison to the input stage noise. The main noise originates therefore from the folded cascode and input stage. MOSFETs that are connected as constant current sources within the folded cascode input stage are ignored as well because of their gates connected to low impedance nodes.

Fig. 4.14a and 4.14b are used to illustrate all significant noise contribution within the NMOS and PMOS input pairs and associated cascode transistors. Both M1/M2

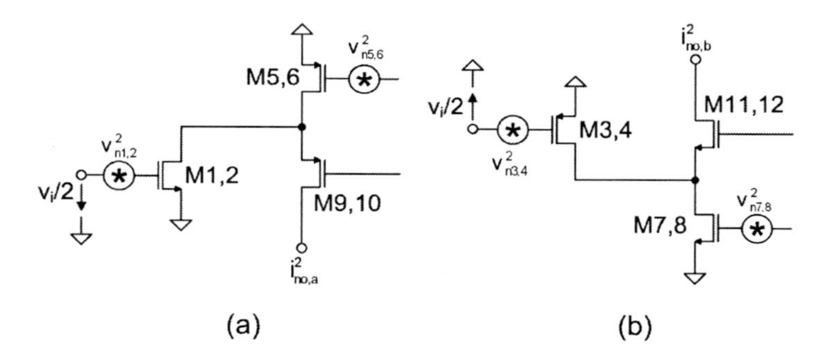

Figure 4.14: Folded-cascode input stage noise sources. (a) NMOS input pair and PMOS cascode transistors. (b) PMOS input pair and NMOS cascode transistors.

and M3/M4 have equal parameters and are connected to identical structures, respectively, which is reflected in Fig. 4.14. The noise analysis therefore treats only one of each symmetric cases. However, a factor of two has to be added afterwards to reflect the contribution of each separate noise source. In addition, the noise from the cascode transistors M5, M6, M11 and M12 is neglected as cascode transistors contribute only negligible noise in cascode or folded-cascode stages [45].

An analysis of the output current noise PSD in Fig. 4.14a gives

$$i_{no,a}^2(f) = g_{m1}^2 v_{n1}^2(f) + g_{m5}^2 v_{n5}^2(f) \tag{4.75}$$

which can be expressed as input-referred dividing $i_{no,a}^2(f)$ by $(4\,g_{m1}^2)$:

$$\begin{aligned} v_{ni,a}^2(f) = \frac{v_{n1}^2(f)}{4} + \frac{g_{m5}^2}{4\,g_{m1}^2} v_{n5}^2(f) \;&=\; \left(\frac{8}{3}kT\frac{1}{g_{m1}} + \frac{KF_n'}{4\,C_{ox}^2 (WL)_1}\right) \\ &+\; \frac{g_{m5}^2}{g_{m1}^2}\left(\frac{2}{3}kT\frac{1}{g_{m5}} + \frac{KF_p'}{4\,C_{ox}^2 (WL)_5}\right) \end{aligned} \tag{4.76}$$

Here, the MOSFET voltage noise sources are replaced by the sum of the MOSFET thermal and 1/f noise as given by (3.42) and (3.48) from Section 3.4. Equation (4.76) can be further simplified which results in

$$v_{ni,a}^2(f) = \left(1 + \frac{g_{m5}}{g_{m1}}\right)\frac{2}{3}kT\frac{1}{g_{m1}} + \frac{1}{4\,C_{ox}^2\,f}\left(\frac{KF_n'}{(WL)_1} + \frac{g_{m5}}{g_{m1}}\frac{KF_p'}{(WL)_5}\right) \tag{4.77}$$

A similar calculation can be conducted for the PMOS folded-cascode input stage in Fig. 4.14b. Its input-referred voltage noise PSD is accordingly given by

$$v_{ni,b}^2(f) = \left(1 + \frac{g_{m7}}{g_{m3}}\right)\frac{2}{3}kT\frac{1}{g_{m3}} + \frac{1}{4\,C_{ox}^2\,f}\left(\frac{KF_n'}{(WL)_3} + \frac{g_{m7}}{g_{m3}}\frac{KF_p'}{(WL)_7}\right) \tag{4.78}$$

In order to obtain the overall input-referred noise all noise contributions have to be added which is two times the noise given by (4.76) and two times the noise of (4.77). Usually, both the input stage transistors and the cascode stage current mirror transistors are designed to have equal input and cascode transconductances, i.e. $g_{m1} = g_{m2...4}$ and $g_{m5} = g_{m6...8}$ hence giving an overall input-referred voltage noise PSD of

$$\begin{aligned} v_{ni,OP}^2(f) &= 2\,v_{ni,a}^2(f) + 2\,v_{ni,b}^2(f) = \\ &\underbrace{\left(1 + \frac{g_{m5}}{g_{m1}}\right)\frac{8}{3}kT\frac{1}{g_{m1}}}_{v_{nith,OP}^2(f)} + \\ &\underbrace{\frac{1}{2\,C_{ox}^2\,f}\left[\frac{KF_n'}{(WL)_1} + \frac{KF_p'}{(WL)_3} + \frac{g_{m5}}{g_{m1}}\left(\frac{KF_p'}{(WL)_7} + \frac{KF_n'}{(WL)_5}\right)\right]}_{v_{nif,OP}^2(f)} \end{aligned} \tag{4.79}$$

where $v_{nith,OP}^2(f)$ and $v_{nif,OP}^2(f)$ indicate the overall thermal and 1/f noise of the op amp. Equating these two expressions and solving for the frequency f gives the corner frequency f_k for the transition between 1/f and thermal noise. The 1/f noise can be suppressed by using chopper modulation as described in Section 4.4.6.

4.2.7 CMRR

Two causes of op amp common-mode gain and hence limited CMRR can be identified for the presented folded-cascode input stage topology [80]. The first one is called random common-mode rejection CMRR_{rd} and is due to mismatch of the input pairs and the second is labeled systematic common-mode rejection CMRR_{st} and arises from asymmetries of the input stage. The common-mode gain due to application of a common-mode voltage at the input pairs is based on the generation of a signal current across the tail current sources I_{diff} in Fig. 4.12 as a results of its finite output impedance G_{Idiff}. This impedance is therefore relevant for both types of CMRR.

The CMRR analysis in [80] considers the mismatch of only one input pair of the rail-to-rail folded-cascode input stage which includes a constant-gm circuitry. This results also holds approximately for both input pairs if the symmetry of the circuit is exploited by assuming that all input, current mirror and cascode transistors have the same transconductance $g_{m,in}$, $g_{m,casc}$, $g_{m,cm}$ and output conductance $g_{ds,in}$, $g_{ds,casc}$, $g_{ds,cm}$, respectively. Hence,

$$\mathrm{CMRR}_{st} = \frac{2\, g_{m,in}\, g_{m,casc}\, g_{m,cm}}{g_{ds,casc}(g_{ds,in} + g_{ds,cm})\, G_{Idiff}} \tag{4.80}$$

and

$$\mathrm{CMRR}_{rd} = \frac{2g_{m,in}}{\frac{\Delta g_{m,in}}{g_{m,in}} G_{Idiff}} \tag{4.81}$$

where $\Delta g_{m,in}/g_{m,in}$ denotes the transconductance mismatch of the input pair. Using the mismatch parameters as described in Section 3.5.1 and considering that the drain current mismatch of a simple differential pair is given by [47]

$$\Delta I_D = g_m \Delta V_{TH} + \frac{\Delta\beta}{\beta} I_D \tag{4.82}$$

the transconductance mismatch can be calculated as

$$\frac{\partial \Delta I_D / \partial V_{gs}}{g_m} = \frac{\Delta g_m}{g_m} = \begin{cases} \frac{\Delta V_{TH}}{V_{OV}} + \frac{\Delta\beta}{\beta} g_m, & \text{for strong inversion} \\ \frac{\Delta V_{TH} g_m}{n U_T} + \frac{\Delta\beta}{\beta}, & \text{for weak inversion} \end{cases} \tag{4.83}$$

where it has been distinguished if the input pair is operation in weak or strong inversion and using the particular g_m definitions from Section 3.3.1. The overall op amp CMRR is dominated by the lower CMRR value, i.e.

$$\frac{1}{\mathrm{CMRR}_{OP}} = \frac{1}{\mathrm{CMRR}_{st}} + \frac{1}{\mathrm{CMRR}_{rd}} \tag{4.84}$$

An op amp specification that is closely related to the CMRR is the op amp input-referred offset voltage as it is also determined by the MOSFET mismatch parameters. A full set of expressions to determine the offset voltage of the rail-to-rail folded-cascode input stage can be found in [78].

4.2.8 Power

The static op amp power dissipation can be determined by summing the currents in all branches of the op amp. These currents are namely the bias currents of the input pairs, two times the floating source current I_{FCS} where it has been assumed that the class-AB control M15/M16 carries the same current like the floating current source, the class-AB control currents I_{ab} and the quiescent current of the output stage, hence

$$I_{OP,stat} = 2\left(I_{diff} + I_{FCS} + I_{ab}\right) + I_q \tag{4.85}$$

The total op amp static power dissipation is given by $V_{DD}\, I_{OP,stat}$. Using (4.51) and (4.56) this can be expressed as

$$P_{OP,stat} = 2V_{DD}\left(I_{diff} + I_{ab}\left(1 + \gamma_{OP} + \frac{\delta_{OP}}{2}\right)\right) \tag{4.86}$$

4.2.9 Programmability

Examining Table 4.1 on page 63 shows that a change in bias current influences the op amp specifications. This in turn opens the possibility to program the specifications of the op amp after fabrication. Of particular interest is the relation between power and bandwidth and the trade-off between power and noise with the latter being most important within the scope of this work. A detailed and systematic methodology on the design of programmable op amps with respect to the noise-power trade-off, in particular using the cascoded Miller compensation topology, was first described by the work of Bronskowski et al. [30].

Two conditions have to be fulfilled to allow a change in the bias current of the Miller compensated op amp:

1. All MOSFETs must not enter the linear region

2. The phase margin should not be altered significantly

The first condition is satisfied by assuring that $V_{DS} > V_{OV}$ for all transistors. The second condition guarantees that the op amp stays stable within the programming range. The bias current dependent parameters of the expression for a sufficient phase margin in (4.74) are mainly the transconductances of the input and output stage g_{mA} and g_{mB}, respectively. The capacitance ratio C_A/C_c is relatively small, it is therefore neglected to ease the derivation of the stability condition. Using the identity

$$\arctan(x) + \arctan(y) = \frac{x+y}{1-xy} \quad (xy < 1) \tag{4.87}$$

in (4.74) and solving the resulting quadratic equation gives

$$\kappa_{PM} = \frac{g_{mA}}{g_{mB}} = \frac{\tan(90° - PM_{stab})(C_L/C_c)}{(C_L/C_c)^2 \tan(90° - PM_{stab}) + (C_L/C_c)} \tag{4.88}$$

Equation (4.88) shows that a constant phase margin is achieved by keeping the ratio of the input and output transconductance, κ_{PM}, constant when changing the bias currents.

The output stage is usually biased in strong inversion whereas for the input stage both weak and strong inversion applies. The use of weak inversion for the input stage allows to have a relative high g_m/I_D ratio as shown in (3.35) which is advantageous with respect to the power-noise trade-off, hence weak inversion is used for the input stage. The relationship between transconductance and drain current for strong and weak inversion has been given by (3.28) and (3.29), respectively. Substituting these in (4.88), assuming that $KP_p(W/L)_{21} = KP_n(W/L)_{22}$ and using (4.56) gives a bias current ratio of

$$\frac{I_{diff}}{\sqrt{I_{ab}}} = \kappa_{PM} n U_T \sqrt{2\delta_{OP} KP_n \left(W/L\right)_{22}} \tag{4.89}$$

There exists one problem to fulfill this relation which is related to the current mirror MOSFETs M5 and M7 that are situated in the cascode. The current in these transistors is the sum of $I_{diff}/2$ and $\gamma_{OP} I_{ab}$. It is therefore impossible to fulfill conditions (4.49), (4.50) and (4.88) simultaneously.

A solution to this problem, which is described in detail in the next section, was presented in [81] where the currents in M5 and M7 depend no longer on both I_{ab} and $I_{diff}/2$, but only on I_{ab}. Assuming that this solution is used, a power-bandwidth relationship and a noise-power relationship can be now derived using the bandwidth, noise and power expressions given above. For this, (4.72) is rearranged with respect to I_{ab} and the value obtained is used to replace I_{ab} in the op amp power equation (4.86) which becomes

$$P_{OP,stat} = V_{DD} \left[I_{diff} + I_{diff}^2 \frac{1 + \gamma_{OP} + \delta_{OP}/2}{2(\kappa_{PM} n U_T)^2 KP_n (W/L)_{22}} \right] \tag{4.90}$$

Now, using (3.29) the transconductance g_{mA} in the op amp unity gain frequency (4.72) is replaced by I_{diff}/nU_T and (4.72) is rearranged with respect to I_{diff}. Using this value for I_{diff} in (4.90) gives an op amp power-bandwidth relationship of

$$P_{OP,stat} = 2V_{DD} \left(n U_T C_c \omega_u + \frac{C_c^2 \left(1 + \gamma_{OP} + \delta_{OP}/2\right) \omega_u^2}{2\kappa_{PM}^2 \delta_{OP} KP_n (W/L)_{22}} \right) \tag{4.91}$$

For the noise-power relationship we consider only thermal noise as 1/f noise is supposed to be removed by chopper modulation as explained in Section 4.4.6. The thermal noise expression in (4.79) includes the transconductances of M5 (M7). Assuming that M5 (M7) carries the current $\gamma_{OP} I_{ab}$, replacing g_{m1} and g_{m5} by their drain current dependent expressions and exploiting the I_{diff}/I_{ab} relation in (4.89) allows to express the g_{m5}/g_{m1} ratio by

$$\epsilon := g_{m5}/g_{m1} = \frac{2}{\kappa_{PM}} \sqrt{\frac{\gamma_{OP}\,(W/L)_5}{\delta_{OP}\,(W/L)_{22}}} \tag{4.92}$$

The power-noise relationship is now given by replacing the g_{m5}/g_{m1} ratio by ϵ in (4.79), replacing the remaining g_{m1} term by its drain current depending expression, rearranging (4.79) with respect to I_{diff} and using this value for I_{diff} in the op amp power equation (4.86), thus

$$P_{OP,stat} = 4V_{DD} \left(\frac{8\,(1+\epsilon)\,kTnU_T}{3\,v_{ni,OP}^2(f)} + \frac{(8\,(1+\epsilon)\,kT)^2\,(1+\gamma_{OP}+\delta_{OP}/2)}{(3\,\kappa_{PM})^2 \delta_{OP} KP_n (W/L)_{22}\,\left(v_{ni,OP}^2(f)\right)^2} \right) \tag{4.93}$$

In sum, changing the bias currents results in two relations regarding the op amp bandwidth and thermal noise with respect to its power dissipation, namely

$$\omega_u \;\propto\; a_1 + \sqrt{a_2 + a_3 P_{OP,stat}} \tag{4.94}$$

$$v_{ni,OP}^2(f) \;\propto\; \frac{b_1}{P_{OP,stat}} + \frac{\sqrt{b_2 + b_3 P_{OP,stat}}}{P_{OP,stat}} \tag{4.95}$$

with $a_1 - a_3$ and $b_1 - b_3$ being the corresponding constants.

4.3 Programmable Op Amps in 350 nm

This section covers briefly the design of programmable op amps in 350 nm technology. Two specific design versions are presented, *OP350a* and *OP350b*, which rely both on the cascoded Miller topology shown in Fig. 4.12. The OP350a version was designed by Dr.-Ing. Carsten Bronskowski [82, 83] and the OP350b version, which is an optimized version of the first one, was designed by Dipl.-Ing. Philipp Meier auf der Heide and was measured and analyzed in [81].

The motivation for describing both versions here has several reasons. At first, OP350a has been used in the AFE of the systems described in Chapter 5. It is therefore essential to know its specifications with respect to the analysis of the front-end. Furthermore, op amp OP350b introduces an additional circuitry which eliminates the problem of the coupled bias currents I_{diff} and I_{ab} through M5 and M6, and M7 and M8. This function is also exploited in the programmable 130 nm op amp of Section 4.4 and its function is therefore of particular interest. Finally, the designs in 350 nm present an opportunity to compare the 350 nm technology to the 130 nm node with respect to challenges in analog design for very small technologies.

4.3.1 Architectures

The circuit topologies of the OP350a and OP350b are shown in Fig. 4.15 where the dotted lines indicate parts that are included in the OP350b only. In addition, the OP350a includes chopper modulation to reduce 1/f noise. Details on this circuit are not given here and can be found in [82, 83]. The OP350a and OP350b designs

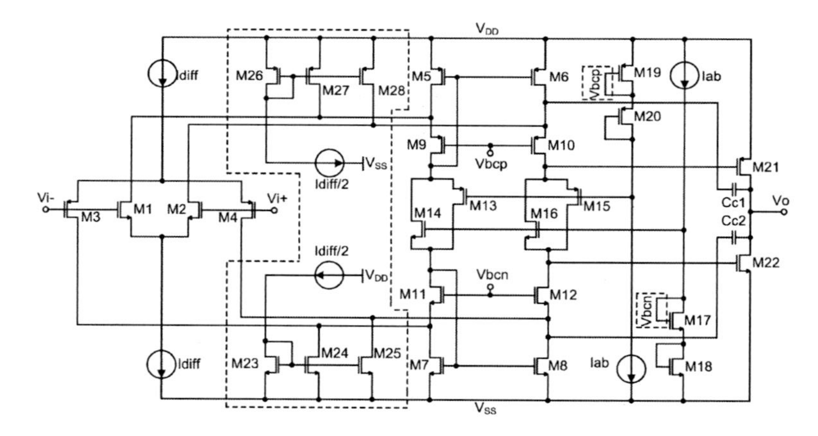

Figure 4.15: OP350a and OP350b (dotted lines) topologies.

resemble the cascoded Miller compensation topology shown in Fig. 4.12 with the OP350b including two modifications to the OP350a. First, the bias voltages for the cascodes (V_{bcn} and V_{bcp}) are generated by the drain voltage of M18 and M19 thus there is no need to add any circuitry to generate these and second, the addition of M23 – M28 which source/sink the bias current of the input pair instead of M5 – M8. The detailed functionality of this additional circuit is given below.

It has to be noted that these op amps do not include a constant-gm circuit. Any change of the transconductance of the input pairs gives rise to distortions as the unity gain frequency $f_u = \omega_u/2\pi$ changes by a factor of two as indicated by (4.72) which affects also the closed loop transfer function. In addition, the op amp is overcompensated when one of the input pairs turns off when considering (4.88) and (4.89), hence too much current is spent in the output stage for these common-mode input ranges. The use of a relative high supply voltage of 3.3 V relieves this situation as the range where both the NMOS and PMOS input stage operate becomes rather large.

4.3.2 Bias Current Decoupling

As indicated in 4.2.9 it is not possible to fulfill the stability condition (4.88) within the programming range because of M5 – M8 sourcing/sinking the bias current of the input pairs $I_{diff}/2$ while being situated simultaneously in the translinear loops M7, M14, M17, M18 and M5, M13, M19, M20. As these translinear loops set the output stages quiescent current I_q, it is not possible to comply with (4.89).

A solution to this problem was presented in the OP350b design [81] by introducing two current mirror structures M23 – M28 between the input pairs and the cascode stage. These current mirrors are biased to source (M27, M28) the quiescent current $I_{diff}/2$ from the NMOS input pair (M1, M2) or to sink (M24, M25) the quiescent current $I_{diff}/2$ from the PMOS input pair (M3, M4). As a result, M5 and M7 no longer source/sink the current $I_{diff}/2$ which decouples the dependence of the associated translinear loop on I_{diff}.

The above arguments hold only for the input pair DC biasing current whereas no change in their AC currents results which is essential for the operation of the circuit. Care has to be taken in the design of the additional current mirrors with regard to their output resistance. It can be seen in Fig. 4.15 that the drains of M24, M25, M27 and M28 are connected in parallel to the drains of M5 – M8 which are part of the input stage output resistance r_A in (4.63). Thus, a too low or changing output resistance of M24, M25, M27 and M28 can lower the op amp gain or introduce distortions in the output signal.

The use of the additional current mirror structure now allows to independently set the bias (or programming) currents I_{diff} and I_{ab} and to fulfill condition (4.88). Resulting from this is a constant phase margin within the programming range.

4.3.3 Realizations and Measurement Results

Both, OP350a and OP350b were realized in a standard CMOS 3.3 V, 350 nm process. The op amps have a size of 1000 μm × 250 μm for the OP350a and 679 μm × 236 μm for the OP350b. The Miller capacitances C_{c1} and C_{c2} were implemented using polysilicon capacitors (poly-caps).

The measurement results are compiled into Table 4.2 to give a quick overview of both designs. The lowest power or lowest noise setups indicate that these op amps are well suited to be used in the IA of a mobile ECG or sensitive EEG system.

An interesting point to note is the open-loop gain at the line frequencies 50/60 Hz because of its influence on the CMRR of the 2-op amp IA (see Section 4.1.2), in particular the value of $\text{CMRR}_{\Delta OP}$ which is given by (4.27). It can be calculated, that the maximum achievable $\text{CMRR}_{\Delta OP}$ value at 50 Hz for an IA using OP350a is 86 dB and 118 dB in low and high power mode, respectively. The same calculation for an IA using the OP350b gives 76 dB and 120 dB. As the lowest CMRR value of the IA sets its overall CMRR, these values limit the achievable CMRR when assuming perfect resistor matching and negligible inter-op amp CMRR mismatch $\text{CMRR}_{\Delta CMRR}$.

Symbol	Parameter	Unit	OP350a		OP350b	
	Power setting		$P_{OP,min}$	$P_{OP,max}$	$P_{OP,min}$	$P_{OP,max}$
V_{DD}	Supply voltage	V	3.3			
P_{OP}	Power dissipation	mW	0.14	29.7	0.27	14
$v_{ni,OP}(f)$	Input noise[1]	$\text{nV}/\sqrt{\text{Hz}}$	14	2	16	3.6
A_{d0}	Open-loop gain	dB	≈ 120 [2]			
f_u	Unity gain freq.[3]	MHz	1.1	39	0.3	48
PM	Phase Margin[3]	°	73	23	≈ 68	
CMRR	CM rejection ratio	dB	82	65	n.a.	
V_{os}	Offset voltage	mV	≤ 0.5 [4]		n.a.	
A_{OP}	Area	mm^2	0.25		0.16	

[1] Considering only thermal noise
[2] Simulation values
[3] For a load of 20 pF (a) and 4 pF || 2 kΩ (b)
[4] Chopper disabled

Table 4.2: Overview of main OP350a and OP350b specifications.

4.4 Programmable Op Amps in 130 nm

This section describes the design and realization of a programmable op amp using a 130 nm standard CMOS technology with 1.2 V supply voltage [84] for use in biomedical signal acquisition systems. The shift towards smaller technologies for biomedical application is motivated twofold. First, it allows to exploit the benefits of additional deep-submicron digital circuits forming together with the analog front-end a mixed-signal system or SoC. Secondly, even though analog circuits do not scale as strong as digital designs with smaller technologies a substantial area reduction is nevertheless expected. Opposed to this are the difficulties in design due to short-channel effects which pose a challenge to be solved.

The topology of the 130 nm programmable op amp is essentially consistent with the 350 nm design what is remarkable because of this topology being supposed to work only with relative high supply voltages (i.e. 2 V) [78]. This has been accomplished using a multi-threshold (multi-V_{th}) process option which became a standard option for newer technologies. Additionally, a constant-gm circuit was added to the input stage to obtain an optimized frequency and distortion behavior. To accomplish this, a novel constant-gm circuitry approach had to be followed as present solutions do not comply with the programmability of the op amp. In addition, a second design is presented which uses additional chopper modulation to suppress 1/f noise. To this day, only a few publications exist describing the design of operational amplifiers in technologies with $L_{min} \leq 130$ nm, these are mainly given by the group of Zimmermann et al. in [85, 86, 71].

4.4.1 Short-Channel and Low-Supply Voltage Design

General Considerations

In contrast to the OP350 approach, problems due to deep submicron effects become visible within the 130 nm node which have to be dealt with for a successful design [87]. One of the main problems using low supply voltages is the reduced number of MOSFETs in the saturation region that can be placed on top of each other. Two kinds of voltages appear in the sum of counting the voltage in one branch of the circuit from 0 V (V_{SS}) to V_{DD}. The first is the drain-source voltage for a device in saturation which is given by $V_{OV} \approx 200$ mV. The second is the gate-source voltage $V_{GS} = V_{OV} - V_{TH}$. Whereas the first cannot be changed, the second includes the threshold voltage which can be influenced by technology. In the present design a standard CMOS 130 nm process with a supply voltage of 1.2 V was used. This technology includes a multi-V_{th} option with 3 standard types of threshold voltage and additionally zero-V_{TH} devices. Using the lowest threshold voltage option increases the minimum supply voltage accordingly. In recent technologies the multi-V_{th} option has become a standard for CMOS processes allowing a general use in analog design.

An (initially undesired) short-channel effect eases even further the design with small supply voltages, namely RSCE which has been described in Section 3.5.3. Although RSCE hinders the design with respect to the MOSFETs bias point setting, it can be exploited to further reduce the threshold voltage as depicted in Fig. 3.11. With the minimal channel length in analog design being approximately three to seven times L_{min}, RSCE poses a benefit to the minimum supply voltage of the circuit.

To calculate whether velocity saturation (see Section 3.2.6) has to be considered in the design the conditions in (3.26) are evaluated for $E_{sat}L$ using L = 130 nm. Using (3.8), (3.9), (3.25), Table 3.2 and assuming $t_{ox} \approx L_{min}/45$ [67] gives a value of $E_{sat}L$ = 507 mV for the n-channel MOSFET and $E_{sat}L$ = 1.9 V in the p-channel case. These values indicate that, although some minor deviation from the square law is expected, no major velocity saturation effects have to be considered.

Finally, gate leakage, mismatch and noise have to be considered. Any large influence of gate leakage current is not expected using the 130 nm technology as outlined in Section 3.5.2. Regarding mismatch no simulation parameters were available for the transistors used, hence preventing any analysis on this topic. As chopper modulation is intended to be used, only the thermal noise is considered. The simulation parameters show no significant deviation from the basic thermal noise equation (3.41) of Section 3.4 indicating that the parameter γ_n has not increased considerably for this CMOS process.

The design was calculated using the basic design equations and approximate parameters extracted from the BSIM models.

Approach using High-Supply Voltage Topology

Considering the folded-cascode stage in Fig. 4.12 with transistors biased in strong inversion limits the minimum supply voltage in the left branch of the cascode to

$$V_{DD} \geq 5V_{OV} \tag{4.96}$$

and

$$V_{DD} \geq 2V_{GS} + V_{OV} \tag{4.97}$$

Equation (4.97) can be rearranged with respect to the maximum threshold voltage $V_{TH,max}$ giving

$$V_{TH,max} \leq \frac{V_{DD} - 3V_{OV}}{2} \tag{4.98}$$

Equation (4.96) defines a minimum supply voltage of 1 V which is ensured by using V_{DD} = 1.2 V whereas (4.98) is only satisfied if the threshold voltage is less or equal to 300 mV. The 130 nm multi-V_{TH} technology used includes 3 standard types of MOSFETs with the lowest threshold voltage option having values of V_{THn} = 0.38 V and V_{THp} = -0.33 V for L_{min}. As an extra option, low-V_{TH} and zero-V_{TH} transistors were available additionally. The use of low-V_{TH} or zero-V_{TH} fulfills condition (4.98). However, if channel lengths of the lowest (standard) threshold option are chosen to be 5 or more times larger than minimum length, the threshold voltages reduce to a value of $V_{THn} \approx V_{THp} \approx$ 250 mV as a results of RSCE. Using these devices with an appropriate channel length satisfies (4.96), hence this solution with standard transistors was used in the design of the 130 nm programmable op amp.

4.4.2 Programmable Op Amp Constant-gm Control

In order to keep the overall transconductance of the rail-to-rail input stage constant with respect to the applied common-mode input voltage a kind of control circuit has to be added. The task of the additional circuit is to increase the current of either the PMOS or NMOS pair if the opposite one turns off hereby increasing its transconductance to a value that compensates for the loss of the overall g_m. Such circuits extend the input stage to a *constant-gm* rail-to-rail one by assuring that $g_{m,n} + g_{m,p} = const.$ with $g_{m,n}$ and $g_{m,p}$ being the transconductance of the NMOS and PMOS input pair, respectively. In the present design the input pairs are biased in weak inversion, i.e. their transconductance is directly proportional to the drain current as given in (3.29). A constant-gm solution suitable for an input stage in weak inversion is the *current switch* approach shown in Fig. 4.16a [45].

The working principle of the current switch is based on the current steering operation of M30. At input common-mode voltages around the middle of the supply voltage range, half of the bias current is directed via M30 to M31 whereas the other half biases the NMOS pair. The current mirror M31/M32 is dimensioned to equalize the difference in parameter n of (3.29) between the NMOS and PMOS pair for equal transconductance, its current ratio is hence given by n_n/n_p. The drain current of M32 is used for biasing the PMOS pair.

At high common-mode voltages the PMOS pair and M30 turn off. The loss of the PMOS pair g_m is counteracted by doubling the NMOS pair current which results in an constant overall transconductance. At low common-mode voltages M30 is fully turned on. The source side of M30 shows a low resistance path in comparison to the NMOS input pair, hence all bias current I_{diff} is directed via M30, M31 and M32 to the PMOS pair hereby doubling its transconductance.

However, this current switch solution cannot be directly used in the programmable op amp design due to the action of the bias current decoupling MOSFETs M24, M25, M27, M28 in Fig. 4.15. The quiescent current of these devices has to be set to zero if the associated input pair turns off to prevent any error in the folded-cascode stage. A modified current switch solution is now presented.

Modified Current Switch Constant-gm Control Topology

To couple the quiescent current of M24/M25 and M27/M28 to the current of NMOS and PMOS input pair the modified current switch solution of Fig. 4.16b has been developed. The operation of the circuit is based on using I_{D30} to control I_{D31} and I_{D34} which set the sourcing/sinking currents of M24, M25, M27 and M28. The current mirror ratio of M23, M24 and M25 have a ratio of 1 : 2, the same holds for M26, M27 and M28. At common-mode voltages around $V_{DD}/2$, M30, M32 and

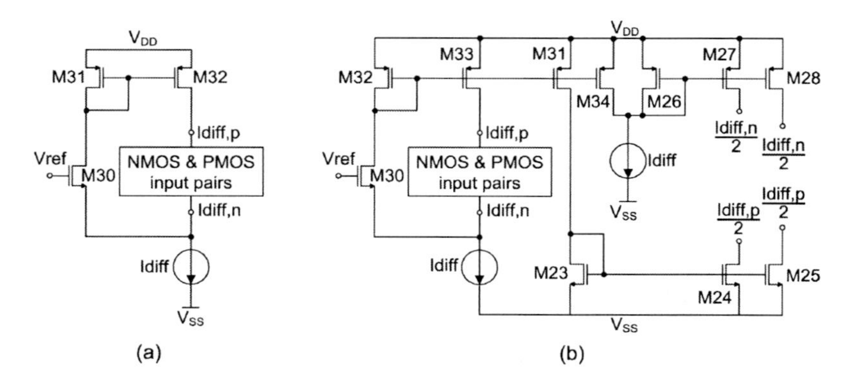

Figure 4.16: Topology of the constant-gm circuits. (a) A conventional current switch. (b) modified version including a programmability extension (the output currents of M24, M25, M27 and M28 are valid only for a medium common-mode voltage).

M33 work as described for the simple current switch circuit. Now, M31 sets I_{D24} and I_{D25} to $I_{diff}/4$, the same holds for M34, I_{D32} and I_{D33} with the exception that I_{diff} is split between M26 and M34.

As in the simple current switch, M30 and the PMOS input pair turn off at high common-mode voltages and all I_{diff} is applied to the NMOS pair to double its transconductance. With I_{D30} being zero M31, M32, M33, and M34 turn off whereas I_{D27} and I_{D28} are set to $I_{diff}/2$ via M26.

For lower common-mode voltages the NMOS input pair shuts off and M30 directs I_{diff} to the current mirror formed by M32 and M33 to double the PMOS input pair gm. Now I_{D24} and I_{D25} are also set by M30 to $I_{diff}/2$ via M23 and M31 and the current of M34 increases significantly. Resulting from this are very low drain current values for M26, M27 and M28.

In the design of the modified current switch constant-gm stage the current ratio of M32/M33 was determined by simulation to be $n_n/n_p \approx 0.94$. The size of M30 was chosen such that at $V_{cm,in} = V_{DD}/2$, I_{D30} equals $I_{diff}/2$ when applying a reference voltage of $V_{ref} \approx V_{DD}/2$. This can be accomplished by using the same gate length for both, M30 and the NMOS input pair and setting W_{30} to $2\,W_1$. As a result, the resistance seen at the source side of M30 and the NMOS input pair becomes equal hence splitting I_{diff}. A simulation of the circuit gives then a value of $V_{ref} = 630$ mV.

The functionality of this constant-gm solution was verified by simulation of the overall g_m of the input pairs as a function of the applied common-mode input voltage for three different op amp power modes as depicted in Fig. 4.17. The total input stage

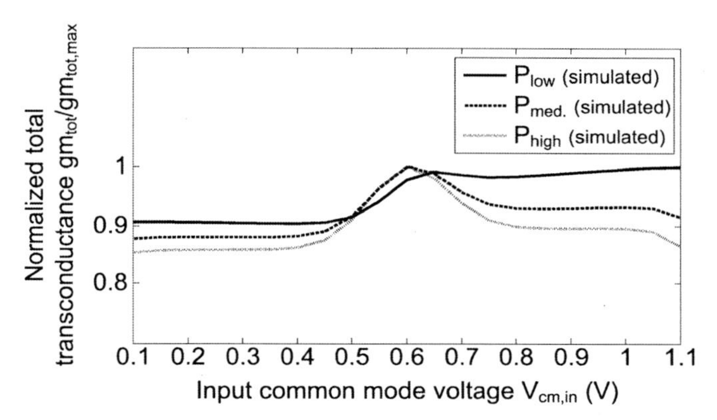

Figure 4.17: Normalized total transconductance of the input stage versus input common-mode voltage.

transconductance $g_{m,tot}$ shows a variation of 9%, 12% and 15% (at P_{low}, P_{medium}, and P_{high}) for a common-mode voltage range of $V_{cm,in} = 100$ mV $- 1.1$ V. The increase in variation is attributed to the input stage going slightly out of weak inversion for higher bias currents. Nevertheless, the functionality of the modified current switch constant-gm circuit is demonstrated by these results.

4.4.3 Design Approach - OP130a

The schematic of the programmable 1.2V 130 nm op amp including the constant-gm circuit is shown in Fig. 4.18. To reduce the design complexity with respect to its frequency response a simple Miller compensation scheme was implemented. The specifications of the first design approach, which will be denoted as OP130a, were targeted to resemble those of the OP350 designs. Of particular interest are the boundaries of the programming range or more precisely the lowest noise and power dissipation within the programming range, respectively, and the phase margin. These were set as the main target specifications in the present design:

T1: A minimum thermal noise of $\sqrt{v^2_{nith,OP}(f)} \approx 2\ldots3\text{nV}/\sqrt{\text{Hz}}$.

T2: The lowest power dissipation should be $P_{OP,stat} \approx 100\mu$W which gives a quiescent supply current of $I_{OP,stat} \approx 90\mu$A at 1.2 V supply voltage .

T3: A phase margin of $PM_{stab} = 72°$ to ensure stability.

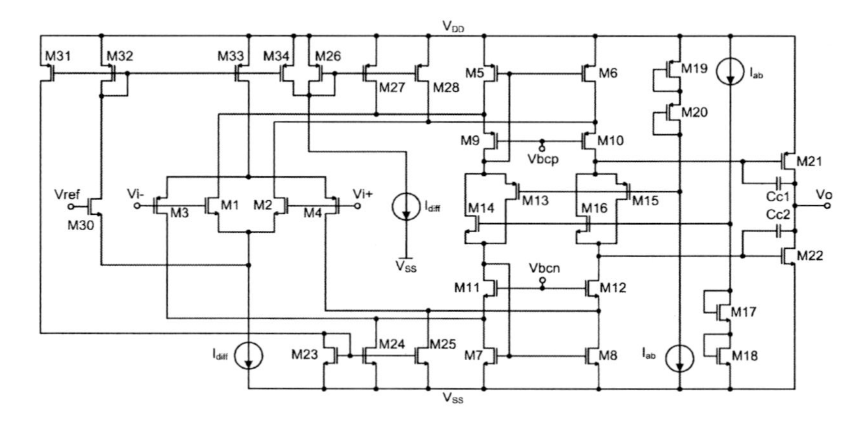

Figure 4.18: Topology of the 130nm op amp including the constant-gm circuit.

The load capacitance and resistance were defined as $C_L = 10$ pF and $R_L = 20$ kΩ. The size of the Miller capacitance was chosen to be 4 times larger than the load capacitance which tends to have a negative influence on parameters like unity gain frequency or slew rate. This size is however needed to limit the current in the output stage for a low-noise setting of the op amp giving otherwise rise to currents in the milliampere range. With the exception of the input and output transistors all channel lengths were fixed to a value of 600 nm thereby setting the threshold voltages to $V_{THn} \approx V_{THp} \approx 250$ mV due to RSCE as explained above. The channel lengths of the input MOSFETs were set to 1 μm in order to obtain a large gate area for 1/f noise reduction. For the channel length of the output stage a value of 360 nm was chosen to achieve a high output g_m without increasing the area consumption excessively. It should be mentioned that the design has generally not been optimized for area consumption as the priority objective was to achieve a functioning 130 nm op amp design.

The first target T1 regarding thermal noise sets the maximum input pair current, a first approximate value is given by the thermal noise term of (4.79) and assuming that $g_{m5}/g_{m1} \ll 1$:

$$
\begin{aligned}
v^2_{nith,OP}(f) &= \frac{8}{3}kT\frac{nU_T}{I_{diff}/2} \overset{!}{\approx} \left(2\ldots 3\ \text{nV}/\sqrt{\text{Hz}}\right)^2 \\
\Rightarrow I_{diff} &= \frac{8}{3}kT\frac{nU_T}{v^2_{nith,OP}(f)/2} \approx 100\ldots 200\,\mu\text{A} \Rightarrow I_{diff,max} = 150\,\mu\text{A} \qquad (4.99)
\end{aligned}
$$

using the highest n value, i.e. $n = 1.5$.

At the other end of the programming range the 90 μA quiescent bias current is divided for the various parts of (4.85), namely:

$$I_{OP,stat} = 2(\underbrace{I_{diff}}_{10\,\mu A} + \underbrace{I_{FCS}}_{1\,\mu A} + \underbrace{I_{ab}}_{1\,\mu A}) + \underbrace{I_q}_{30\,\mu A} + \underbrace{I_{Biasnetwork}}_{30\,\mu A} = 84\,\mu\text{A} \tag{4.100}$$

with the output stage transistors exhibiting a three times larger bias current than each of the input stages in order to achieve a high g_{mB}/g_{mA} ratio needed for a sufficient phase margin. The choice of 1 μA for both branches of the cascode stage and I_{ab} simplifies the design with respect to keep the associated transistors from going into weak inversion. The current ratios given set the corresponding parameters of (4.51) and (4.56) to $\gamma_{OP} = 1$ and $\delta_{OP} = 30$.

Having defined the main target specifications T1 – T3 and some initial parameter values allows to design the programmable operational amplifier using the design equations of Section 4.2, Fig. 4.19 illustrates the main dependencies. Considering

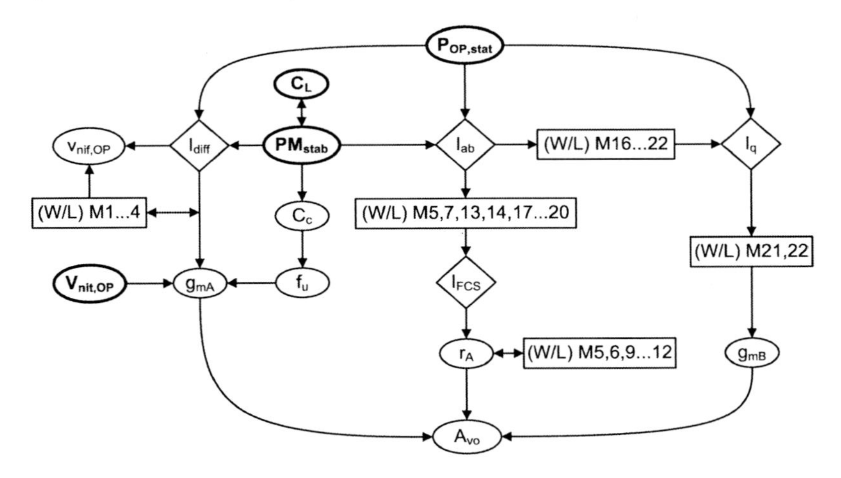

Figure 4.19: Dependencies of the OP130 design parameters. The bold symbols indicate input parameters.

that basic equations are used implies that only coarse design parameters will be obtained as a starting point even when generating a set of SPICE parameters for different MOSFET sizes and biasing. Therefore, appropriate values have to be determined afterwards by simulation.

The phase margin condition (4.88) plays a crucial role to enable the programmability. Using the given values of $C_L = 10$ pF and $C_c = 40$ pF and targeting for a phase margin of $PM_{stab} = 72°$ to compensate for neglecting C_A in (4.88) gives

$$\kappa_{PM} = \frac{g_{mA}}{g_{mB}} \approx 0.42 \tag{4.101}$$

These results can be used to calculate the (W/L) ratio of M21 and M22 by inserting into (4.89) and rearranging, thus

$$(W/L)_{22} = \frac{(I_{diff})^2}{2\delta_{OP} KP_n I_{ab} (\kappa_{PM} n U_T)^2} \approx 47 \ldots 105 \tag{4.102}$$

for $n = 1 \ldots 1.5$ and a similar analysis for the PMOS output transistors yields $(W/L)_{21} \approx 168 \ldots 379$.

The MOSFETs geometries within translinear loops of the floating current source and the class-AB control are set using relations (4.49), (4.50), (4.53) and (4.54). Considering the translinear loops of just the NMOS devices (as symmetric arguments hold for the translinear loops of the PMOS devices) reveals several free variables. A first constraint is given by setting $(W/L)_{14} = (W/L)_{16}$ to simplify the design. The same argument applies to M17 and M18, thus $(W/L)_{17} = (W/L)_{18}$. In addition, it was considered that at the lowest noise setting or equally highest bias current a voltage "headroom" of approximate $2\,V_{OV} \approx 300 \ldots 400$ mV is left to the current sources I_{ab}. This sets the sum of the maximum gate-source voltages $V_{GS,max}$ of M17 and M18 to 800...900 mV, respectively. The highest bias current of M17/M18 can be calculated from (4.89) and assuming $I_{diff,max} = 150\,\mu$A which gives $I_{ab,max} \approx 50\,\mu$A. This value is actually lower due to the input pair moving slightly from weak to strong inversion for higher bias currents. Neglecting this effect to obtain a first coarse value gives the sizes of M17/M18 by inserting $I_{ab,max}$ into (3.12) and rearranging with respect to (W/L), thus

$$\frac{2\,I_{ab,max}}{KP_n (V_{GS,max} - V_{TH})^2} \approx 4.8 \ldots 8.5 \tag{4.103}$$

which also determines the (W/L) ratio of M7, M14 and M16. Having defined M7 sets also the size of M8 as both form a current mirror with ratio one. The same procedure is used for the PMOS devices which gives the sizes for M5, M6, M13, M15, M19 and M20.

The folded cascode transistors M11 and M12 were sized (with $(W/L)_{11} = (W/L)_{12}$ for symmetry reasons) by trading off their transconductances against their capacitances which both increase for a higher (W/L) ratios assuming a fixed channel length. The capacitance of M12 contributes significantly to the value of C_A in (4.74) whereas a higher g_{m12} increases the output resistance of the first stage r_A and hence the op amp open-loop gain A_{vo}. Using simulation a good compromise was found by setting $(W/L)_{12} = 2$. The same procedure applies also to the NMOS cascode transistors M9 and M10. For the cascode stage the two bias voltages V_{bcp} and V_{bcn} have to be set to such values, that saturation of all related MOSFETs is assured. This was implemented by generating these voltages across additional drain connected NMOS and PMOS devices which are not shown in Fig. 4.18 (these devices were biased by I_{ab}, similar to the OP350b solution in Fig. 4.15). A direct generation of the bias voltage at the drains of M17 and M20 could not be implemented with the given geometries.

Finally, the NMOS and PMOS input pairs have to be sized to be in weak inversion for the whole programming range. The MOSFETs turn from strong to weak inversion at $V_{GS} = (V_{GS})_{ws}$ which is given by (3.20). Using the maximum drain current of M1 ($I_{D1,max} = 75\,\mu$A) and determining its (W/L) ratio (assuming its V_{GS} is slightly larger than $(V_{GS})_{ws}$ so (3.12) can still be used) gives the inequality

$$(W/L)_1 = (W/L)_2 > \frac{2\,I_{D1,max}}{KP_n(2nU_T)^2} \approx 51\ldots115 \qquad (4.104)$$

to drive M1 and M2 into weak inversion. For the PMOS pair a ratio of $(W/L)_3 = (W/L)_4 \approx 183\ldots414$ is obtained using the same calculation. In addition, it has to be verified that the 1/f noise corner frequency lies within a reasonable value as it will be determined by the area of the input pair devices as shown in (4.79). A very high corner frequency implies that $(W\,L)$ has to be increased while keeping the (W/L) ratio constant. For the present design with the channel lengths of the input pairs set to 1 μm, the bias depending 1/f noise corner frequency varies between $10 - 107$ kHz (see Table 4.7).

Using the hand calculations as initial parameters, appropriate values were obtained by simulation. The exact values of I_{ab} and I_{diff} normalized to I_{ab} and bias voltages for the programming range are given in Table 4.3. Table 4.4 shows the exact (W/L) ratios for all MOSFETs.

I_{ab}	I_{diff}	V_{bcp}	V_{bcn}	Comment
1 μA	13.4 μA	770 mV	420 mV	$P_{OP,low}$ - Low-power mode
5 μA	46.4 μA	640 mV	550 mV	
10 μA	72.8 μA	570 mV	620 mV	
15 μA	93.1 μA	520 mV	670 mV	$P_{OP,med}$ - Medium-power mode
20 μA	110.7 μA	480 mV	710 mV	
25 μA	127.3 μA	450 mV	740 mV	
30 μA	143.3 μA	420 mV	770 mV	$P_{OP,high}$ - High-power mode

Table 4.3: Programming setup.

MOSFET	(W/L)	MOSFET	(W/L)
M1	(55.77/1)	M12	(1.2/0.6)
M2	(55.77/1)	M13	(20.5/0.6)
M3	(324.15/1)	M14	(3.2/0.6)
M4	(324.15/1)	M15	(20.5/0.6)
M5	(6.07/0.6)	M16	(3.2/0.6)
M6	(6.07/0.6)	M17	(2.3/0.6)
M7	(1.21/0.6)	M18	(2.3/0.6)
M8	(1.21/0.6)	M19	(13/0.6)
M9	(6.07/0.6)	M20	(13/0.6)
M10	(6.07/0.6)	M21	(116.4/0.36)
M11	(1.2/0.36)	M22	(23.38/0.36)

Table 4.4: OP130a MOSFET geometries (μm).

4.4.4 Realization - OP130a

The op amp realized in the standard CMOS 1.2 V, 130 nm process uses the standardly available triple-well option to suppress the body effect of the NMOS devices. The fabricated op amp in an overlay picture (due to dummy metalization) of a chip micrograph and a layout plot from the design tool is shown in Fig. 4.20. For the Miller compensation, metal-insulator-metal (MiM) capacitors are used. They are situated in the topmost metalization level and were placed aside the op amp to eventually allow a view of the active area. The capacitors at the bottom of Fig. 4.20 are additional MiM capacitors to stabilize the supply voltage.

A placement above the active area of the op amp would reduce the area consumption (assuming no coupling effects). Including the capacitors, the op amp has an area of 0.109 mm^2. By placing the Miller capacitors above of the op amp only half of this area is needed.

4.4.5 Results - OP130a

The design of the 130 nm op amp test setup poses a challenge with respect to the low supply voltage of 1.2 V. Difficulties arise from the low number of available components for this supply voltage and from the standard measurement equipment which is usually designed for higher operating voltages. Additionally, the setup becomes more sensitive to power supply spikes and ESD effects.

The OP130a was measured using the setups given in Appendix B with the exception of open-loop gain. This was accomplished by zeroing the offset voltage and applying a very small and low-frequency input sine to an open-loops setup. The

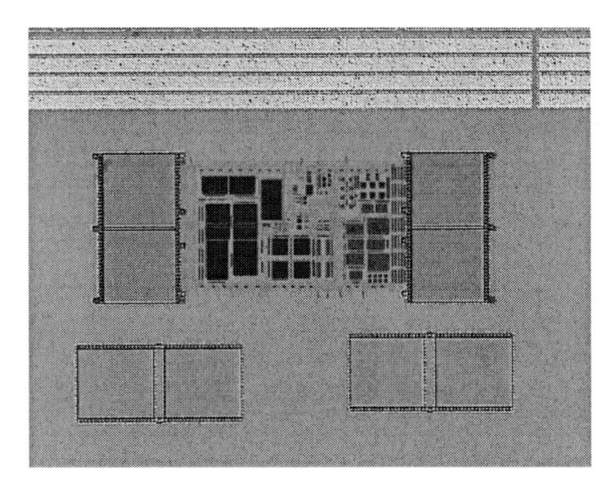

Figure 4.20: Overlay of chip micrograph and layout showing the OP130a.

results of this (rather error prone) procedure have been verified by a subsequent open-loop gain measurement using the techniques of Appendix B. The parameters input-referred offset, PSRR and CMRR were not included in the first measurement setup. These were determined for the second design which is described in Section 4.4.6.

Programmability and Phase Margin

The programmability of the OP130a was tested by setting the bias currents given in Table 4.3. At the lowest power mode of the programmability range a power dissipation of $P_{OP,stat} \approx 200\mu\text{W}$ was measured. This value includes the power dissipation of the op amp which was targeted to be approximately $100\mu\text{W}$ in Section 4.4.3, the constant-gm circuit and the power dissipation of additional circuitry to generate the bias voltages and currents. The value lies hence within the approximated target specification of the power dissipation.

The input-referred thermal noise versus power dissipation trade-off and the unity-gain versus power dissipation trade-off are shown both in Fig. 4.21. The minimum thermal noise at the highest bias current setup was determined at a frequency of 200 kHz to suppress the influence of 1/f noise. The measured value of $\sqrt{v_{nit,OP}^2} = 5.1\text{nV}/\sqrt{\text{Hz}}$ is approximately two times larger than the initially targeted value. The increase is caused by three different influences, first the g_{m5}/g_{m1} ratio in (4.79) cannot be fully neglected, second a lower g_{mA} at the highest bias current due to M1 to M4 going slightly out of weak inversion and third a small influence of 1/f noise. Considering this, the measured values match satisfyingly the specifications given.

When the power consumption is reduced, the noise increases as described by the proportionality given in (4.94). Likewise, the unity-gain frequency follows a change in bias current or power dissipation as given in (4.95). The noise value for the lowest-power mode could not be measured reliably because of the reduced bandwidth.

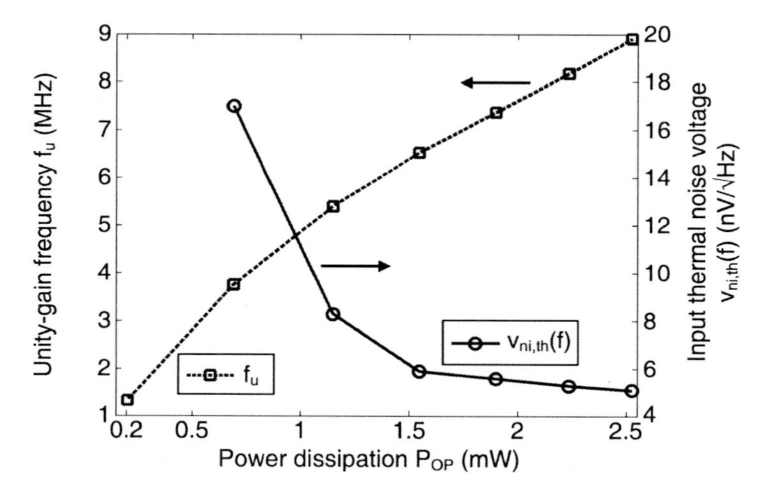

Figure 4.21: OP130a noise versus power and unity-gain frequency with load conditions of $R_L = 22\ \text{k}\Omega$ and $C_L = 13.5$ pF.

The measured phase margin within the programming range varies from 58° in the low-power mode to 62° in the high power mode, Table 4.5 shows the values for each programming step. The mean value is $\overline{PM_{stab}} = 59.4°$ with a maximum deviation of 2.8°. These values approve the design approach and assumptions made in Section 4.4.3 and assure stability within the whole programming range while simultaneously limiting the ringing of the output voltage when applying a voltage step at the input. Additionally, for all programming settings, stability was verified using a voltage follower setup.

Const-gm

To verify experimentally the simulation results of Fig. 4.17, the constant-gm stage was tested by sweeping the input common-mode voltage and measuring the unity-

I_{ab} (μA)	1	5	10	15	20	25	30
Comment	$P_{OP,low}$			$P_{OP,med}$			$P_{OP,high}$
Phase margin PM	62.3°	59.7°	59°	59°	59.2°	58.5°	57.8°

Table 4.5: Measured OP130a phase margin within programming range with load conditions of $R_L = 22$ kΩ and $C_L = 13.5$ pF.

gain frequency. The unity-gain frequency is linearly related to the transconductance of the input pair assuming a constant Miller capacitance as given by (4.72), hence this value can be used to indirectly determine any g_{mA} variation. Care has to be taken when the output stage reaches the minimum or maximum output voltages $V_{out,min}$ and $V_{out,max}$ as given in (4.58) and (4.59). At this point, the output MOSFETs turn into the linear region thereby making relation (4.72) invalid.

The variation of the measured values differs significantly from the simulation of Section 4.4.2 as shown in Fig. 4.22. The variation from maximum unity-gain frequency in low-power mode for $V_{cm,in} = 200$ mV to $V_{DD}-200$ mV is approximately 34%. An analysis of these results shows that in the technology used the output resistance of the input transistors decreases by more than one decade for drain-source voltages lower than 400 mV. In addition, the output transistors move into the linear region for output voltages close to the supply rails. The first effect leads to a degradation of the functionality of the constant-gm circuit whereas the second effect masks the effect of the constant-gm circuit when using the unity-gain frequency technique. The functionality degradation can only be improved by moving the input pair further into the weak inversion regime or by developing a circuitry to keep the drain to source voltage of the input pairs within a stable range. In order to derive more accurate values of the input stage overall transconductance and omit the effect of the output stage it would be advisable to realize just the input and cascode stages within a test chip.

General Parameters

In addition to the main programmability parameters of concern additional op amp specifications were measured within the programming range. Table 4.6 shows an overview of these for the low, medium and high power dissipation setups labeled by $P_{OP,low}$, $P_{OP,med}$ and $P_{OP,high}$. The open-loop gain is ≈ 80 dB for the low and medium power setup and decreases slightly at the high power setup. The total harmonic distortion of about -70 dB proves that no serious distortions are added to the circuit using the additional transistors M24/25 and M27/28 to allow the programmability. The input-referred offset stays relatively constant at an acceptable value within the whole programming range whereas the slew rate increases by one order of magnitude

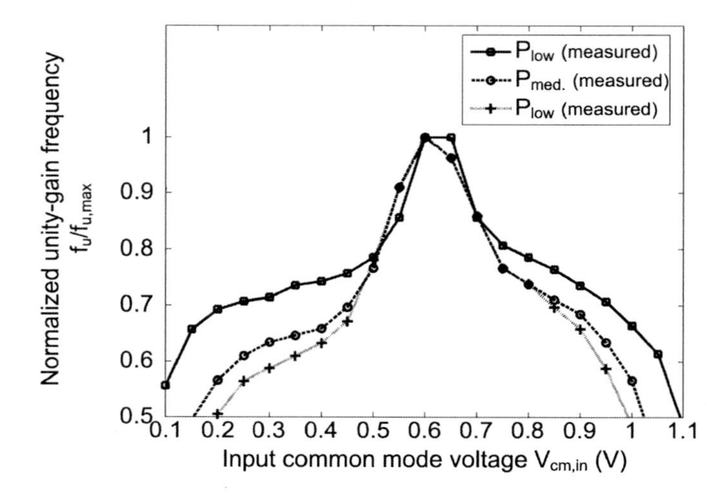

Figure 4.22: Measured normalized unity-gain frequency versus input common-mode voltage.

when sweeping the bias current from the lowest to the highest power setup. This was indeed expected and can be explained by the slew rate dependence on the bias current of the input pairs. Finally, the maximum output current sets the minimum load resistance to a value of 128 Ω.

4.4.6 Chopper Stabilization Version - OP130b

The limitations of reducing op amp imperfections by means of device sizing and layout techniques give rise to the need of additional circuit techniques. These include most notably the reduction of amplifier offset and noise using methods like autozeroing or chopper stabilization [88]. Within the scope of CMOS circuits for biomedical signal acquisition 1/f noise plays a major role because of both biomedical signals and 1/f noise sharing approximately the same bandwidth. Only a significant reduction of 1/f noise makes it possible to acquire biomedical signals precisely using analog CMOS circuits. An approach to reduce 1/f noise was tested in [89, 90] using a modified OP350a. The method exploits physical effects on the device level to lower the 1/f noise of a single MOSFET. The noise reduction obtained is however significantly smaller than using chopper stabilization which was used for the second 130 nm design.

Symbol	Parameter	Unit			
	Power setting		$P_{OP,low}$	$P_{OP,med}$	$P_{OP,high}$
V_{DD}	Supply voltage	V	1.2		
P_{OP}	Power dissipation	mW	0.2	1.5	2.5
$v_{ni,OP}(f)$	Input noise[1]	nV/$\sqrt{\text{Hz}}$	n.a.	5.9	5.1
A_{d0}	Open-loop gain	dB	79	79	76.5
f_u	Unity gain freq.[2]	MHz	1.3	6.5	8.9
PM	Phase Margin[2]	°	62	59	58
V_{os}	Offset voltage	mV	0.2	0.3	0.3
THD	Total harmonic dist.[3]	dB	-70	-72	-69
SR	Slew rate	μV/s	0.1	1	1.5
$I_{out,max}$	Max. I_{out} (sink/source)	mA	4.7/8.3	4.7/8.3	4.7/8.3
A_{OP}	Area (excluding caps)	mm^2	0.052		

[1] Considering only thermal noise @200 kHz
[2] For a load of 13.5 pF || 22 kΩ
[3] Using an 1 kHz input sign of 1 Vpp applied to an voltage follower configuration

Table 4.6: Measured OP130a specifications.

This section presents a 130 nm op amp design called OP130b which includes chopper stabilization to minimize the 1/f noise. The design is identical to the OP130a except for the additional chopper circuitry which is mostly based on [83].

Short Introduction to Chopper Stabilization

Chopper stabilization is based on transposing the input signal to a higher frequency by means of modulation before 1/f noise is added. A following demodulation step moves the signal back to the baseband whereas 1/f noise is transposed to the higher frequency and can be low-pass filtered afterwards. Figure 4.23 illustrates the basic steps:

1. The input signal v_i is multiplied by a square wave at frequency f_{ch} and its PSD is accordingly transposed to the chopping frequency f_{ch}.

2. The input-referred op amp noise, most notably 1/f noise, is added.

3. Both, the signal and the noise are amplified by the first gain stage.

4. Applying a second multiplication on the first gain stage output signal by a square wave at frequency f_{ch} reconstructs the original signal within the baseband and moves the low frequency noise to the chopping frequency.

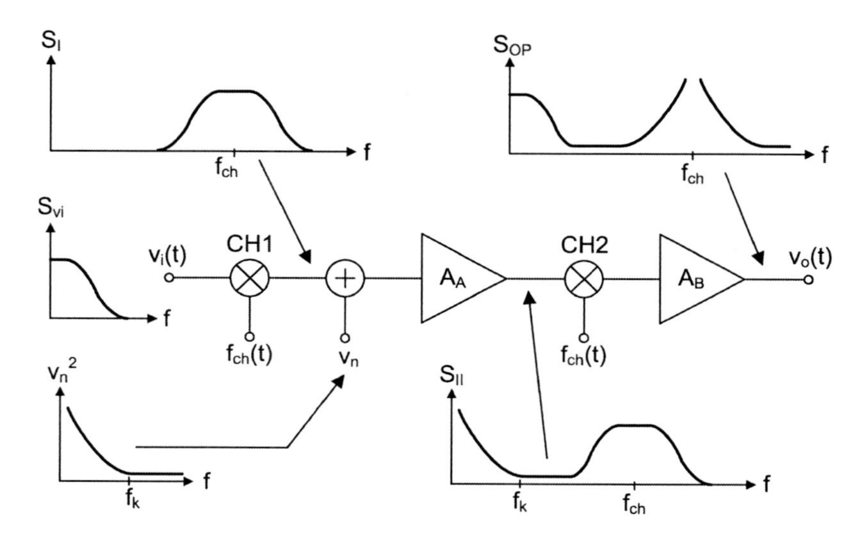

Figure 4.23: Basic principle of chopper stabilization applied to the op amp to reduce 1/f noise.

The effect of second chopper CH2 on the noise $v_n^2(f)$ was analyzed in [88] and results in an op amp output noise of

$$v_{no,OP}^2(f) = \left(\frac{2}{\pi}\right)^2 \sum_{n=-\infty}^{+\infty} \frac{1}{n^2} v_n^2 \left(f - n f_{ch}\right), \; n \text{ is odd} \tag{4.105}$$

The effect of chopper stabilizing results in a thermal noise floor which equals approximately the sum of the original thermal noise and some fraction of the 1/f noise within the baseband which can be approximated as additional thermal noise for $|f \cdot f_{ch}| \leq 0.5$ and $f_u \cdot f_k \gg 1$ [88]:

$$v_{ni,OP}^2(f) \approx v_{nit,OP}^2(f) \left(1 + 0.8525 f_k / f_{ch}\right) \tag{4.106}$$

In addition to the reduction of 1/f noise, chopper stabilization also reduces the input-referred offset of the op amp. Ideally, modulating the input-referred offset by the square wave of CH2 would result in a zero offset due to transposing the DC offset to the chopper frequency. Nevertheless, there exists some offset due to non-idealities of the switches which are caused by switching spikes as a result of clock feedthrough and charge injection. The switching activity of the modulator results in succeeding positive and negative spikes which are amplified and demodulated, i.e.

after demodulation all spikes are positive which results in a DC value after low-pass filtering. However, this offset voltage is usually significantly smaller in comparison to the original op amp offset

Chopper Stabilization Design

The chopper stabilization design follows the principle depicted in Fig. 4.23, i.e. the modulator is placed before the input stage with MOSFETs M1-M4 and the demodulator is situated in front of the output stage consisting of M21 and M22. Fig. 4.24 shows the implementation of this scheme based on [83]. The circuit of the modulator is straight forward. Using two non-overlapping clocks Φ_1 and Φ_2, the input V_{in+} is connected to M2/M4 and V_{in-} to M1/M3 for phase Φ_1. Within phase Φ_2, V_{in+} is connected to M1/M3 and V_{in-} to M2/4. Details on the generation of non-overlapping clocks are not discussed here but can be found in [53].

The demodulator circuit was incorporated into the cascode stage and consists of two identical parts for the NMOS and PMOS section within the cascode. Its operation will be just explained for the NMOS part. For the clock phase Φ_1 the circuit connection are identical to those of the original circuit for the current mirror connections of the M7/M8 gates to the drain of M11. This also holds for the class-AB control for M22 with its gate connected to M15/M16. At phase Φ_2 the input signals are swapped by the modulator which has to be reversed by the demodulator. This is achieved by swapping the functionality of the two branches within the cascode stage connecting the gates of M7/M8 to the drain of M12 and using M13/M14 for class-AB control of M22. A precondition to allow this demodulator implementation is a fully symmetric cascode stage with respect to all transistor geometries and bias voltages which is the case for the OP130a design.

The modulator and demodulator switches were implemented using CMOS transmission gates (see Fig. 4.25) with identical PMOS and NMOS geometries to counteract the effect of charge injection. The modulator devices M1-M4 have a W/L ratio of $24\mu m/0.12\mu m$ and for the demodulator devices a W/L ratio of $4\mu m/0.12\mu m$ was chosen. The larger device width for the modulator is due to reduction of noise which is not as much critical for the demodulator, hence the smaller channel width.

Choosing the chopper frequency f_{ch} to be equal to the 1/f corner frequency f_k results in a good trade-off between 1/f noise reduction and charge injection spikes. The 1/f corner frequency changes within the programming range according to (4.79) and approximate 1/f corner frequencies for all bias setups are given in Table 4.7.

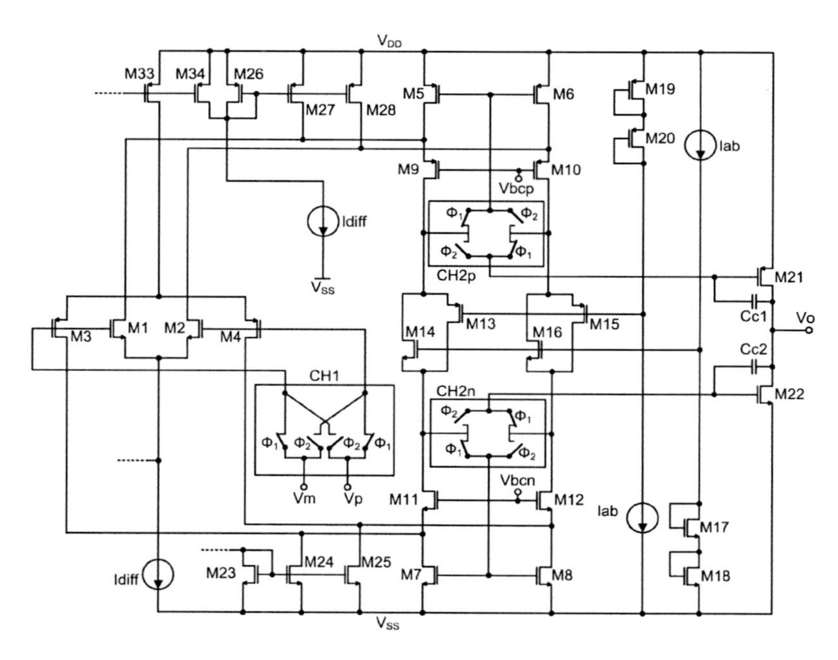

Figure 4.24: Topology of the OP130b including choppers with ideal switches, for better clarity only parts of the constant-gm circuit are shown.

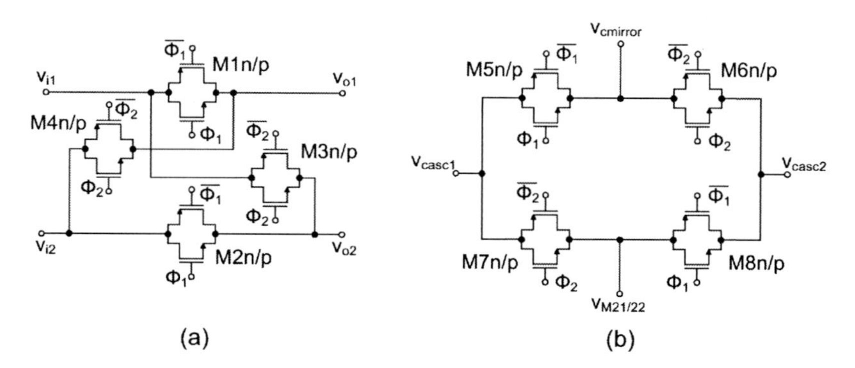

Figure 4.25: Implementation of the chopper stabilization switches. (a) modulator; (b) demodulator.

	$P_{OP,low}$			$P_{OP,med}$			$P_{OP,high}$
I_{ab} (μA)	1	5	10	15	20	25	30
f_k	10 kHz	30 kHz	52 kHz	63 kHz	81 kHz	89 kHz	107 kHz

Table 4.7: Approximate 1/f noise corner frequencies within the programming range.

4.4.7 Realization and Results - OP130b

The OP130b op amp was realized using the same 130 nm technology like the OP130a design. In addition to the op amp a small digital circuit was added to generate the non-overlapping clocks Φ_1 and Φ_2. This circuit allows to select digitally between the two clock *dead times* of 1.7 ns and 3.2 ns. The MiM Miller compensation capacitors were placed above the active area of the op amp, Fig. 4.26 shows the overlay picture of the chip micrograph and the layout plot from the design tool. The overall chopper stabilized op amp area is 328 μm $\times$ 206 μm (317 μm $\times$ 206 μm excluding the non-overlapping clock generation circuit).

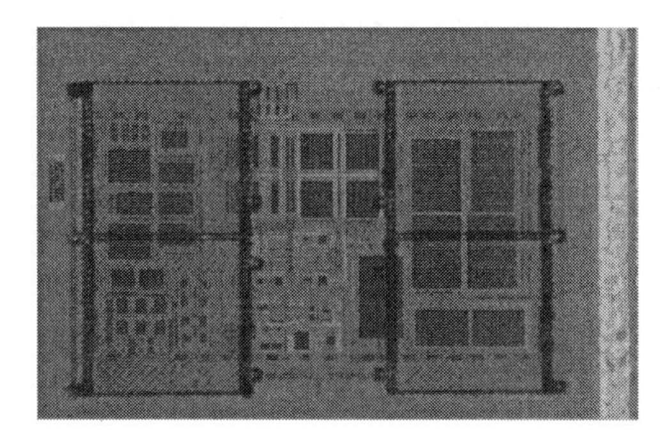

Figure 4.26: Overlay of chip micrograph and layout showing the OP130a.

Noise, CMRR and Offset

The chopper stabilized op amp noise was measured using the highest bias current setup. The chopper frequency f_{ch} was set to 100 kHz which is approximately equal to the 1/f noise corner frequency at this bias setup. According to (4.106), the theoretical input-referred noise floor should increase from the OP130a thermal noise PSD of 5.1 nV/$\sqrt{\text{Hz}}$ in lowest noise setup to a value of approximately 7 nV/$\sqrt{\text{Hz}}$. The measurement results of the output noise divided by the gain of the op amp are shown in Fig. 4.27. A slight increase of noise at lower frequencies with chopper enabled is visible in Fig. 4.27. This increase is supposed to result from 1/f noise of the output stage which is situated after the demodulation step. The peaks at multiple integers of 50 Hz are due to power line interferences coupling into the test setup. The mean value of the noise with chopper enabled (excluding power line interference) was determined to be $v_{ni,OP}(f) \approx 7.1\mathrm{nV}/\sqrt{\mathrm{Hz}}$ and agrees to the theoretical value given above.

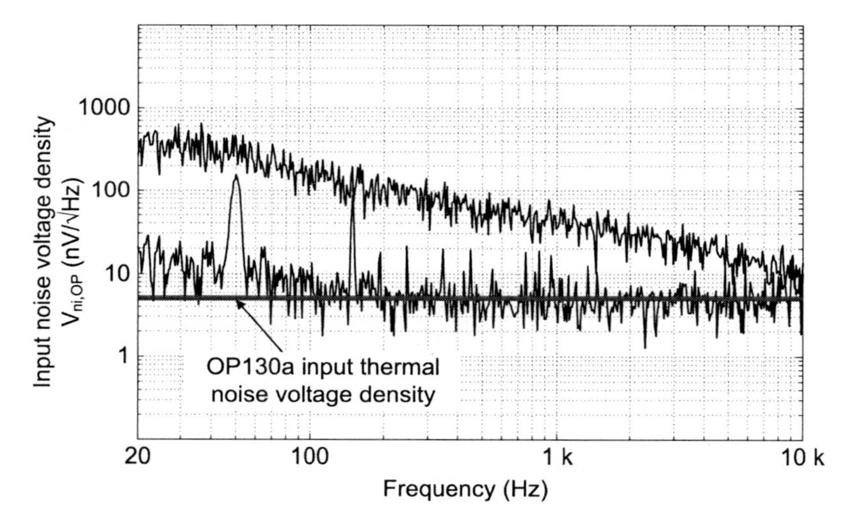

Figure 4.27: Input-referred noise of the OP130b using a chopper frequency of 100 kHz.

The op amp DC common-mode rejection ratio CMRR_{OP} was determined with chopper enabled by measuring the offset voltage difference when shifting the signal ground reference to +300 mV and -300 mV [76], see Appendix B. This measurement was performed for each of the programming range bias setups, Fig. 4.28 shows the CMRR versus power dissipation results. The CMRR varies accordingly between a value of 72 dB and 64 dB for the lowest and highest bias setup respectively. Considering (4.80) and (4.81), this result can be explained generally by an output resistance reduction of the op amp transistors at high bias currents which outperforms the concurrent increase in transconductance. This effect accounts also for the reduction of DC open loop gain at high bias currents as given in Table 4.6.

The measured input-referred offset V_{os} with chopper disabled has a maximum value of 600 μV. When the chopper is activated using a chopper frequency of $f_{ch} =$

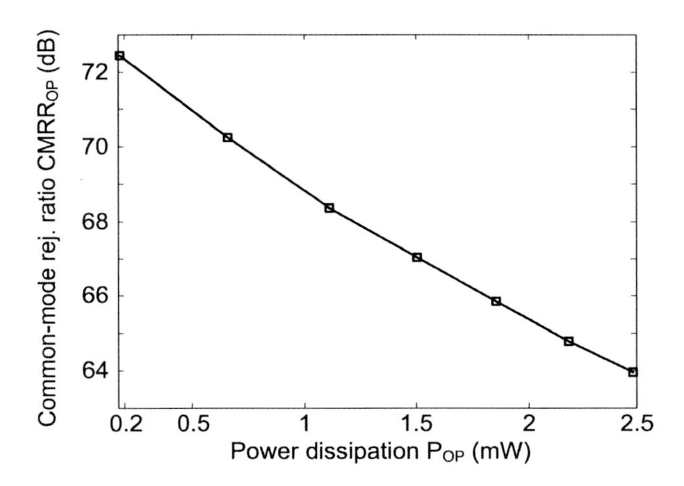

Figure 4.28: OP130b common-mode rejection ratio versus power consumption with chopping enabled ($f_{ch} = 100$ kHz).

100 kHz the input-referred offset reduces to a value of $V_{os} \leq 36\mu$V for an input common-mode voltage at signal ground.

Supplemental Parameters

Additional parameters like open-loop gain A_{d0} and static power dissipation P_{OP} with disabled chopper were measured and are essentially consistent with the OP130a design. By enabling the chopper, the DC open-loop gain drops by approximately 1dB from the value with disabled chopper stabilization. In addition to the op amp, the power dissipation of the digital circuit to drive the chopper switches has to be considered when operating. The corresponding value was simulated to be less than 200 nW for a chopper frequency of 100 kHz which is negligible in comparison the op amp power dissipation.

As a last missing parameter, the power supply rejection ratio (PSRR) measures the ability of the op amp to reject power supply voltage changes. It is defined separately for the positive and negative supply voltages (V_{DD} and V_{SS}) as

$$\mathrm{PSRR}_{+} = \frac{A_{d0}}{v_o/v_{DD}} \tag{4.107}$$

and

$$\mathrm{PSRR}_{-} = \frac{A_{d0}}{v_o/v_{SS}} \tag{4.108}$$

with A_{d0} being the usual open-loop gain, v_{DD} and v_{SS} are small signal voltages of the respective power supplies and v_o is the op amp output voltage response to these. The PSRR value given above is normally expressed in dB. Figure 4.29 shows the DC PSRR measured using the test setup that is described in [76] (see also Appendix B). It can be seen, that changes in the positive supply voltage are generally better rejected than those of the negative supply which have a minimum value of ≈ 70 dB.

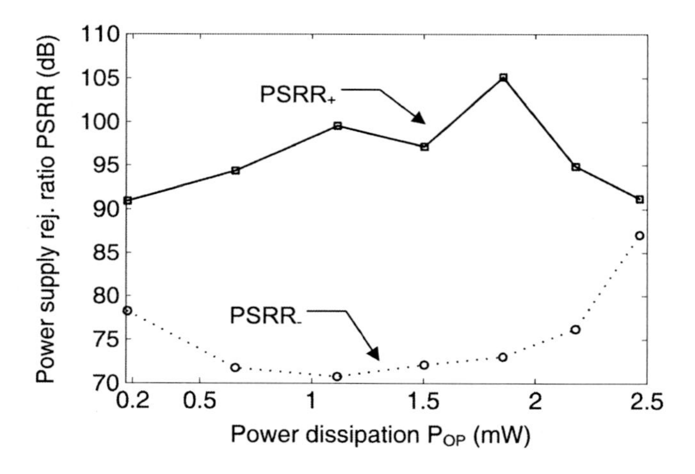

Figure 4.29: OP130b power supply rejection ratio versus power consumption with chopper disabled.

4.5 Application Examples

In this section two application examples using the programmable op amps for the design of the functional blocks (see Section 4.1) will be given. The first example describes a 130 nm 2-op amp IA design and the second an implementation of a Postamp/PGA using 130 nm technology. Application and design examples for the 350 nm op amps are not given here as these are extensively described in the design of the biomedical signal acquisition systems in Sections 5.2 and 5.3.

4.5.1 IA130 - A 130nm 2-Op Amp IA Approach

In order to estimate the performance of the 130 nm programmable op amps as a preamplifier within a biomedical acquisition channel, a 2-op amp IA using the OP130a design was simulated. The choice of using the version without chopper stabilization is based on the difficulties to simulate the effect of chopper modulation. Except for the input-referred noise and offset, all other main op amp parameters are nearly identical for the OP130a and OP130b designs. The parameters considered here are all given in the descriptions of the ideal 2-op amp IA in Section 4.1.2.

The 2-op amp IA using the OP130a design will be referred to as *IA130* and is depicted in Fig. 4.30. The circuit uses a supply voltage of $V_{DD} = 1.2$ V. To compare this IA to the 350 nm design of Section 5.2, the 130 nm 2-op amp IA design was set to have the same gain using the same gain setting resistor values, i.e. $A_{IA130} = 20$ and $R_4 = R_1 = 7.6$ kΩ and $R_3 = R_2 = 400\,\Omega$.

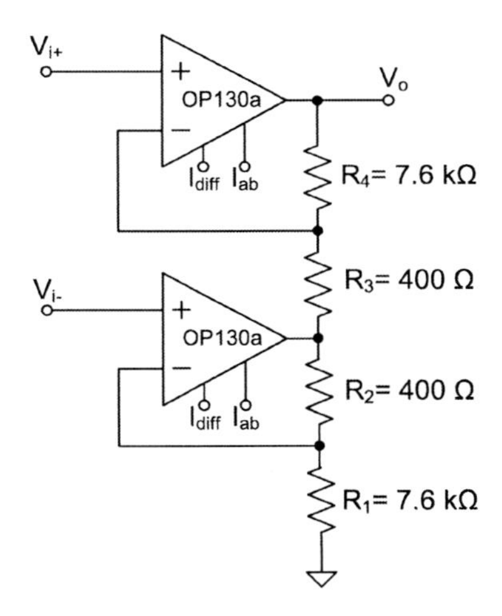

Figure 4.30: IA130 circuit showing the 130nm 2-op amp IA Approach.

The gain error ΔA_{IA130} is not considered here as it depends on the resistors geometries which were taken ideal for this simulation. The theoretical input common-mode range of (4.23) gives an ICMR of ± 0.57 V. A simulation of the input common-mode range gives a value of ICMR $= -0.54$ V$\ldots+0.52$ V which is slightly smaller than the theoretical value. This minor reduction is due to the OP130a output stage going slightly out of saturation for output voltages near the power rails.

The IA130 power dissipation is listed in Table 4.8 for seven OP130a bias current settings. The power dissipation varies between 368 μW and 5.53 mW for the lowest and highest bias current settings, respectively. These values are similar to those of the IA in Section 4.1.2 and show that the 130 nm circuit operating at 1.2 V is comparable to the 350 nm 2-op amp IA with respect to power dissipation.

P_{IA130}	I_{ab}	I_{diff}	Comment
368 μW	1 μA	13.4 μA	$P_{IA130,low}$ - Low-power mode
1.4 mW	5 μA	46.4 μA	
2.38mW	10 μA	72.8 μA	
3.23 mW	15 μA	93.1 μA	$P_{IA130,med}$ - Medium-power mode
4.03 mW	20 μA	110.7 μA	
4.79 mW	25 μA	127.3 μA	
5.53 mW	30 μA	143.3 μA	$P_{IA130,high}$ - High-power mode

Table 4.8: 130IA static power dissipation and bias currents within programming range.

The IA130 noise was simulated at low, medium and high bias current setup for the EEG and EEG bandwidth and is given in Table 4.9. The value $v^2_{ni,IA130}(f)$ in the first column gives the IA130 input-referred noise PSD including thermal and 1/f noise and $v^2_{ni,IA130}(f)^*$ in the second column presents the input-referred noise PSD excluding the 1/f noise of the OP130a input and cascode stage to approximate the influence of chopper stabilization. The noise value was additionally multiplied by $\sqrt{2}$ to model the influence of 1/f noise to the thermal noise floor as given by (4.106). It can be seen that the version without chopper stabilization does not match the EEG and ECG noise specification which are given by $max\{v_{ni,IA130}(f)\}_{EEG} \approx 1\,\mu$Vpp and $max\{v_{ni,IA130}(f)\}_{ECG} \approx 5\,\mu$Vpp for the EEG and ECG bandwidth, respectively due to the high 1/f noise component.

The values approximating the use of chopper stabilization for the highest bias setup and the EEG bandwidth are a factor 1.27 lower than calculating the 2-op amp noise by using the measured OP130b noise data for the ideal 2-op amp noise calculation given by (4.32). This shows the limitations of ideally simulating the noise of the IA and excluding the 1/f noise of the OP130a input and cascode stage. However, even if the input-referred RMS noise will increase by a factor of up to 1.95 for EEG at highest bias setup or 2.95 for ECG at lowest bias setup in an implemented system indicates that such a version would meet the noise specifications. In addition, the use of chopper stabilization would not significantly increase the power consumption of the op amp and additionally decrease the input-referred op amp offset. The only drawback of using chopper stabilization is a more complex circuitry

$v^2_{ni,IA130}(f)^{1,2}$	$v^2_{ni,IA130}(f)^{*1,3}$	Bandwidth	Power mode
16.8 μVpp	1.06 μVpp	0.5 – 100 Hz (EEG)	$P_{IA130,low}$
21.6 μVpp	1.69 μVp	0.05 – 250 Hz (ECG)	$P_{IA130,low}$
19.2 μVpp	558 nVpp	0.5 – 100 Hz (EEG)	$P_{IA130,med}$
24.6 μVpp	881 nVpp	0.05 – 250 Hz (ECG)	$P_{IA130,med}$
21 μVpp	521 nVpp	0.5 – 100 Hz (EEG)	$P_{IA130,high}$
27 μVpp	808 nVpp	0.05 – 250 Hz (ECG)	$P_{IA130,high}$

[1] The peak-to-peak voltage noise Vpp is calculated as the 6-fold value of the root-mean-square voltage noise Vrms
[2] Input-referred noise including thermal and 1/f noise.
[3] Input-referred noise excluding the 1/f noise of the OP130a input and cascode stage.

Table 4.9: 130IA input-referred noise for EEG and ECG bandwidths and for selected power setting.

because of applying the chopper clock from either outside the chip or generating it internally.

Furthermore, the 130 nm 2-op amp was simulated with respect to CMRR limitation due to finite open-loop gain of the op amps used. In general, the CMRR value of the 2-op amp IA is given by the three components $\text{CMRR}_{\Delta OP}$, $\text{CMRR}_{\Delta R}$ and $\text{CMRR}_{\Delta CMRR}$ of (4.30) in Section 4.1.2. However, the analysis was limited to $\text{CMRR}_{\Delta OP}$ because of both, resistor and inter-op amp CMRR mismatch being layout dependent which has not been realized for this analysis. In addition, the resistor mismatch can be minimized by calibration techniques as explained in Chapter 5.

The simulated $\text{CMRR}_{\Delta OP}$ at 50 Hz versus the 2-op amp IA power consumption results are shown in Fig. 4.31. It can be clearly seen that the $\text{CMRR}_{\Delta OP}$ is limited to values around 80 dB which agrees to the theoretical value using (4.27) when considering that simulated open-loop gain values are used which differ slightly from the DC open-loop gain values in Table 4.6.

Finally, the NEF value given in (2.9) was calculated using the above results to compare this design to state of the art biomedical signal acquisition amplifiers. The NEF values for the power setups $P_{IA130,low}$, $P_{IA130,med}$, and $P_{IA130,high}$ and EEG bandwidth are NEF = 12.2, 19, and 23.2, respectively. These values lie within the medium NEF trend line (NEF = 10 . . . 20) for systems published from the year 2000 up to this day [91] which makes this design compatible to systems using conservative technologies.

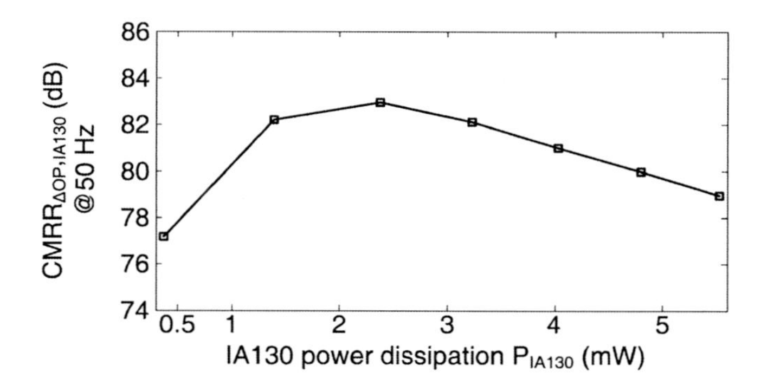

Figure 4.31: IA130 common-mode rejection ratio $\text{CMRR}_{\Delta OP,IA130}$ due to the finite OP130a open-loop gain at 50 Hz.

4.5.2 MyoC1 PGA

A PGA/postamp using the OP130a design was developed for an implantable system to acquire EMG signals from inside the human body to control a bionic hand [92]. The overall system consists of multiple signal acquisition channels, each sharing a single ADC by means of multiplexing. The PGA is supposed to be situated behind the multiplexing circuit and should both increase the channels gain as well as drive the input circuitry of the ADC properly. The PGA is based on the general design described in Section 4.1.3.

Design Realization

The MyoC1 PGA topology resembles the one of Fig. 4.9 where the tapping of the resistor ratio value was realized using CMOS transmission gates and inverters to generate the negated control signals SW_1 - SW_3 as shown in Fig. 4.32. By digitally controlling the transmission gates $\text{TG}_0 - \text{TG}_3$ the gain can be set to be either 1, 2.4, 12 or 24. Each transmission gate consists of a parallel connected NMOS and PMOS transistor with geometries of $W/L = 1\,\mu\text{m}/0.16\,\mu\text{m}$ and $W/L = 3\,\mu\text{m}/0.17\,\mu\text{m}$. These sizes present an optimal value with respect to the on and off resistances of the transmission gate. The feedback resistor values $R_{1a} = 140\text{ k}\Omega$, $R_{1b} = 80\text{ k}\Omega$ $R_{1c} = R_2 = 10\text{ k}\Omega$ were chosen to approximately match the noise of the ADC as given in (2.7) while limiting the dynamic power consumption. The layout area of the MyoC1 PGA does not differ from the OP130a design which is 0.109 mm^2 because all feedback resistors and switches were placed below the Miller capacitors as shown in the design tool plot in Fig. 4.33.

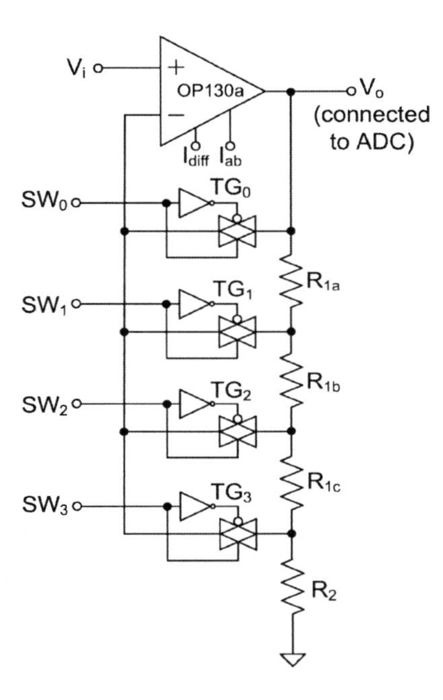

Figure 4.32: MyoC1 PGA topology using CMOS transmission gates to set the gain.

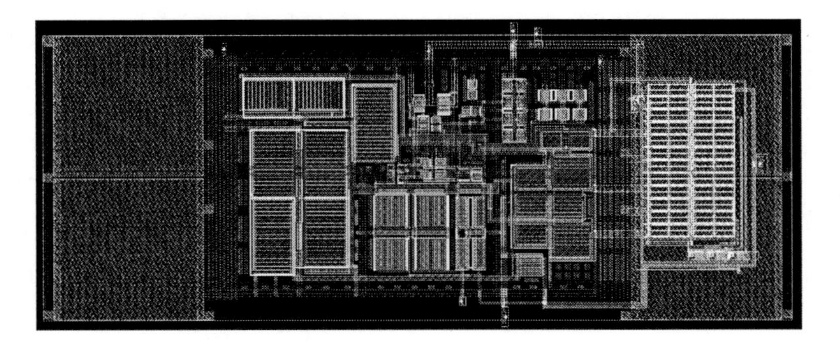

Figure 4.33: MyoC1 PGA layout generated from the design tool.

Performance

The performance of the circuit is given by means of simulation results because of the circuit still being in production at the time of writing. Most importantly, the noise performance for two, system specified bandwidths and the power dissipation of the PGA are listed in Table 4.10 and 4.11. With the OP130a being programmable, all specifications are given for three power setting modes $P_{PGA,low}$, $P_{PGA,med}$ and $P_{PGA,high}$ which correspond respectively to the $P_{OP,low}$, $P_{OP,med}$ and $P_{OP,high}$ bias current settings of Table 4.3.

PGA Gain	1	2.4	12	24
@ $P_{PGA,low}$	Input-referred noise[1]			
Bandwidth 100-800 Hz	1.29 μV	1.53 μV	1.37 μV	1.33 μV
Bandwidth 100-1500 Hz	1.49 μV	1.89 μV	1.62 μV	1.56 μV
@ $P_{PGA,med}$	Input-referred noise[1]			
Bandwidth 100-800 Hz	1.51 μV	1.71 μV	1.57 μV	1.54 μV
Bandwidth 100-1500 Hz	1.73 μV	2.08 μV	1.84 μV	1.78 μV
@ $P_{PGA,high}$	Input-referred noise[1]			
Bandwidth 100-800 Hz	1.6 μV	1.8 μV	1.66 μV	1.63 μV
Bandwidth 100-1500 Hz	1.83 μV	2.17 μV	1.94 μV	1.89 μV

[1] Input-referred RMS voltage noise (Vrms)

Table 4.10: MyoC1 PGA noise specification (simulation results).

	Power dissipation	
@ $P_{PGA,low}$	Static	Dynamic @ 200 kHz
	152 μW	295.6 μW
@ $P_{PGA,med}$	Static	Dynamic @ 300 kHz
	1.29 mW	1.52 mW
@ $P_{PGA,high}$	Static	Dynamic @ 300 kHz
	2.06 mW	2.456 mW

Table 4.11: MyoC1 PGA power dissipation specification (simulation results for the PGA set to unity gain).

The increased noise for higher op amp bias currents originates from a slightly higher 1/f noise at higher bias currents (ideally, the 1/f noise should be constant as given by (4.79)). With the OP130a having no chopper stabilization, 1/f noise becomes the main noise contribution within the specified bandwidth. Therefore it would be advisable to use a lower power dissipation setup to reduce the input-referred noise. However, even at high noise setups the input-referred RMS noise of the PGA is more than ten times smaller than one half of LSB of the 1.2V, 10-bit ADC. In addition, for high frequency channel multiplexing it becomes necessary to have a fast output stage to charge or discharge the input capacitance of the ADC.

To quantify the settling time of the PGA an equivalent circuit of the MyoC1 system circuitry at its input and output was used to simulate the approximate timings, Fig. 4.34 depicts the simulation setup. At the beginning, the transmission gates TG_1 and TG_2 are turned off and the input signal V_i is applied to TG_1. Next, TG_1 turns on whereas TG_2 stays turned off and the PGA starts to set V_o to the voltage that corresponds to the gain setting. After 3 μs, a value which is determined by the overall system, TG_2 turns on and connects the PGA output to the capacitor $C_L = 70$ pF which serves as a model of the ADC input stage. This scheme is used to obtain worst case timings values for both rising and falling edges by setting, for each gain, the input voltage V_i to a value that corresponds to an PGA output voltage of ± 540 mV. The settling times are defined by the time at which the PGA output voltage error is less than 1/2 LSB ($= 586\,\mu$V) of the following 10-bit ADC. The results are listed in Table 4.12 and show that this circuit can handle (gain dependent) channel multiplexing frequencies of 300 kHz $-$ 1.2 MHz at highest power setup $P_{PGA,high}$.

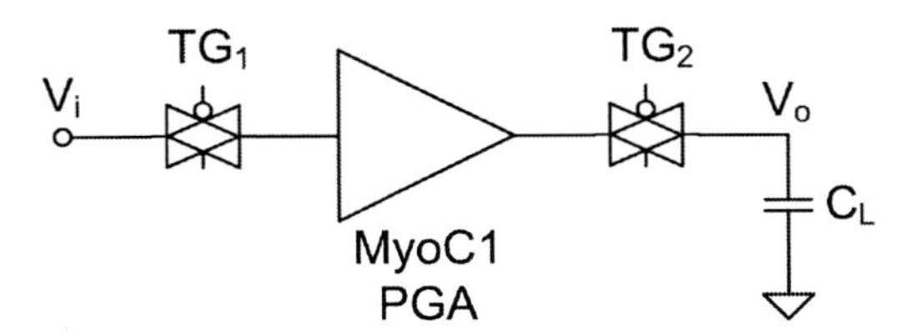

Figure 4.34: Circuit to simulate the MyoC1 PGA timings.

4.6 Conclusion

This chapter has described the implementation of analog front-end building blocks, in particular the design of programmable operational amplifiers which are being used in the AFE.

With the CMRR of the instrumentation amplifier being one of the most important

PGA Gain	1		2.4		12		24	
Edge[1]@ $P_{PGA,low}$	r	f	r	f	r	f	r	f
Settling time[2](μs)	15	10.7	n.a.[3]	7.73	n.a.[3]	n.a.[3]	> n.a.[3]	n.a.[3]
Edge[1]@ $P_{PGA,med}$	r	f	r	f	r	f	r	f
Settling time[2](μs)	0.762	0.464	1.81	2.36	1.73	1.74	2.2	1.31
Edge[1]@ $P_{PGA,high}$	r	f	r	f	r	f	r	f
Settling time[2](μs)	0.510	0.234	1.2	2.25	1.26	1.2	0.827	0.18

[1] r: Rising edge, f: Falling edge

[2] Time needed for an error $< 1/2$ LSB

[3] No settling of output voltage within given timeframe

Table 4.12: MyoC1 PGA channel multiplexing timings (simulation results).

parameters for biomedical signal acquisition, a detailed analysis has been given. The results show that the CMRR does not only depend on the external resistor matching, but also on the CMRR matching between the two op amps being used within the instrumentation amplifier and on the (frequency dependent) op amp open-loop gain.

Comparing the op amp specifications of the 350 nm and 130 nm designs shows that it is possible to use newer technologies with lower supply voltages to achieve similar performance. On the one hand, the use of the smaller 130 nm technology is however complicated due to short-channel effects and inaccurate hand-modeling. On the other hand, this drawback is balanced by the possibility to integrate more complex digital circuitry within the same chip. This makes the use of 130 nm or even smaller technologies an attractive alternative to the commonly used *conservative technologies* for biomedical circuits. Considering the noise and power parameters of the front-end, it has been shown that programmable operational amplifiers allow to use just one design for either low-noise mode applications like EEG or low-power mode like mobile ECG.

Chapter 5

System Solutions

The functional analog front-end block implementations of the previous chapter were used in the development of two CMOS integrated systems for biomedical signal acquisition. The systems include the aforementioned front-ends, analog-to-digital conversion, digital interfacing circuitry and optionally on-chip digital signal processing capabilities. This chapter starts with an introduction giving background information on the project in which these systems were developed including basic system design specifications. In Section 5.2, a 3-channel integrated system solution including a digital signal processor (DSP) is presented and Section 5.3 details the development of a 10-channel system. Conclusions are finally given in Section 5.4.

5.1 Introduction

The work presented here was accomplished within the framework of an European Union (EU) project titled *Multi Monitoring Medical Chip for Homecare Applications (M3-C)* [93]. The goal of this two and a half year lasting project was the design, simulation, fabrication and testing of a universal integrated circuit for the acquisition of biomedical signals. Companies and universities from five European countries were involved in this project with the *Hamburg University of Technology* and *K.U. Leuven* (Belgium) being responsible for the circuit design part. The basic requirement specifications of the M3C system-on-chip (SoC) include

- Applicability to a wide range of biomedical signal types including ECG, EEG, EP and EMG
- Several channels with each one including an instrumentation amplifier, a programmable gain amplifier, low- and high pass filters and ADCs
- Autocalibration for gain accuracy and CMRR as well as DC-offset suppression

- Digital interfacing for output of the measured data and control of the system settings
- Additionally:
  - Built-in self test (BIST)
  - Respiration measurement
  - Digital signal processor

Three integrated circuits using a CMOS 3.3 V, 350 nm technology were realized over the course of this project, chronologically named *M3C1*, *M3C2* and *M3C3*. The M3C1 contains several system components like AFE, DSP or respiration circuitry placed separately on the chip for testing purposes, a chip micrograph of the M3C1 is shown in Fig. 5.1[1] The M3C2 and M3C3 are fully functional systems and will be presented in the next two chapters.

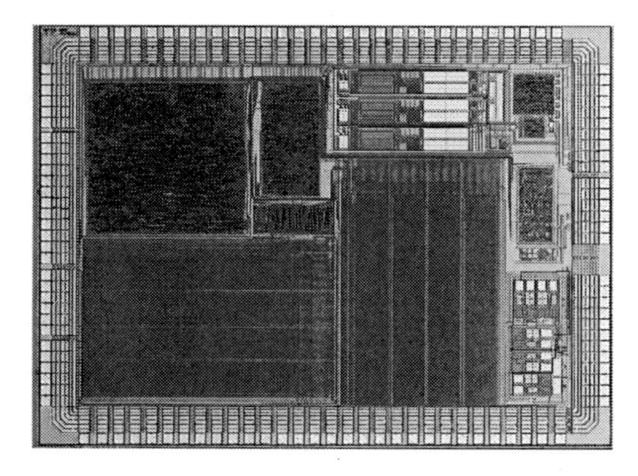

Figure 5.1: Chip photograph of the first M3C test/evaluation IC (M3C1).

5.2 M3C2 System Solution

5.2.1 Introduction and System overview

The M3C2 is a 3-channel biomedical signal acquisition SoC solution prototype [94, 95, 96] and was designed jointly by the author and Dipl.-Ing. Wjatscheslaw Galjan who in particular implemented the silicon version of the DSP. In addition,

[1] The book cover of reference [45] shows actually the top center part of the M3C1.

the calibration, test signal generation and built-in self test circuitry were provided by Ir. Nick Van Helleputte and the 2nd-order $\Sigma\Delta$-ADC was implemented by Dr.-Ing. Alexander Mora-Sanchez. Fig. 5.2 depicts the system architecture. Each channel consists of an AFE, a $\Sigma\Delta$-ADC consisting of a $\Sigma\Delta$-modulator and a decimation filter and a digital I/O which is composed of three UARTs (Universal asynchronous receiver/transmitter) and 16 general purpose I/O pins. The internal DSP can be used for a wide range of on-chip signal processing, e.g. signal feature detection or signal compression. A timing and control unit (TCU) is used to control the setup of the system, e.g. its gain or bandwidth, and generates the internal clock signals needed for the system. The built-in self test and autocalibration unit is used to calibrate the CMRR and gain accuracy of the front-end. For this purpose, test signals are generated on-chip by a stimulus generator. Finally, a respiration circuitry was added for concurrently measuring the patient's respiration and performing an ECG recording.

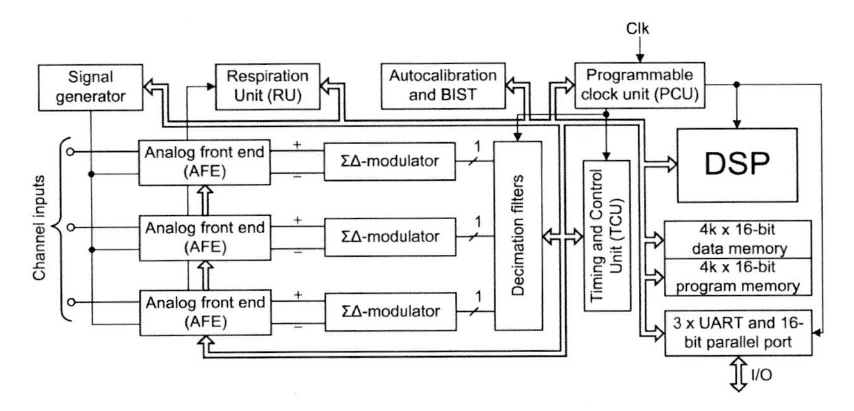

Figure 5.2: System architecture.

5.2.2 Analog Front-End

The M3C2 analog front-end is essentially designed like the type C front-end in Section 2.5 using a prototype DC-suppression circuit. Therefore, a detailed description of the circuit is given by assuming that the DC-suppression has been turned off, the DC-suppression circuitry will be briefly described in the next Section. Fig. 5.3 shows the basic structure of the M3C2 analog front-end which is programmable with respect to gain, noise-power trade-off and low-pass filter cut-off frequency. The IA input terminals V_{i+}, V_{i-} are used in the usual manner and V_{sgnd} is connected to signal ground. This input scheme will change if the DC-suppression circuitry is used. The front-end design specifications, based on the biomedical amplifier consideration of Section 2.4, are given in Table 5.1.

Parameter	Condition/Remarks	Value
Signal input range	Ext. programmable	$\pm$ 5 mV
		$\pm$ 20 mV
		$\pm$ 80 mV
Total input-referred	0.05 Hz - 250 Hz (ECG)	$<$ 5 μVpp
voltage noise[1]	0.5 Hz - 70 Hz (EEG)	$<$ 1.0 μVpp
	0.01 Hz - 5 kHz (EMG)	$<$ 6.0 μVpp
	0.1 Hz - 3 kHz (EP)	$<$ 2.2 μVpp
CMRR	@ 50 Hz	$>$ 63 dB
Gain error	Between channels	$<$ 1%
Power dissipation	ECG	7 mW per channel
including ADC	EEG	25 mW per channel

[1] The peak-to-peak voltage noise Vpp is calculated as the 6-fold value of the root-mean-square voltage noise Vrms

Table 5.1: Design specifications for the M3C2 analog front-end.

For the front-end, in particular the IA, the OP350a programmable operational amplifier of Section 4.3 is used with an external, variable bias current of $I_{bias,low} = 2\,\mu$A to $I_{bias,high} = 20\,\mu$A, hence the ability of the front-end to be either in a low-power dissipation mode or in a low-noise mode. In addition to the programmable gain of the PGA of 1 or 4, the IA was realized to be also programmable with respect to two gain factors, namely 20 and 80. The gain of the PGA of 4 is set using the feedback resistors $R_1 = 30$ kΩ, $R_2 = 10$ kΩ and connecting the op amp output to the next stage using switch SW_1. A unity gain configuration is obtained by directly connecting the output voltage of the IA to the next stage, thereby disconnecting the op amps output from the signal path. As a result, the front-end can be programmed to have a gain of 20, 80 or 320. These values correspond to a maximum input-referred signal range of $\pm$82.5 mV, $\pm$20.63 mV, and $\pm$5.16 mV considering the supply voltage of $V_{DD} = 3.3$ V.

The anti-aliasing low-pass filter can be programmed to a cut-off frequency of 2.4 kHz or 12.5 kHz and was realized by a Sallen-Key topology, see Section 4.1.4. Finally, the signal is converted from single-ended to differential because of the differential input of the $\Sigma\Delta$-modulator.

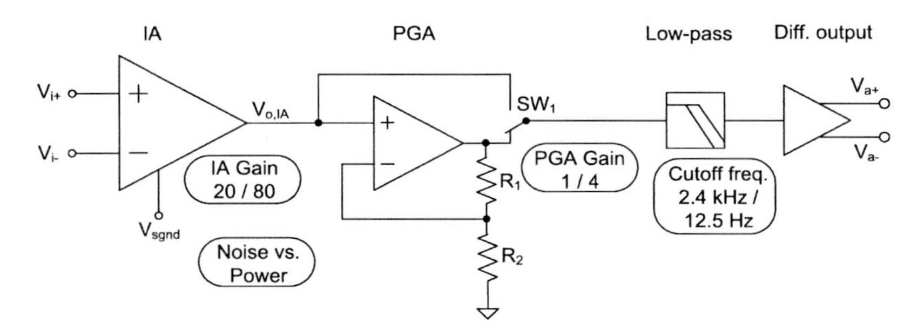

Figure 5.3: Detailed view of the M3C2 front-end. The rounded rectangles show parameters for the programmability of the front-end.

Instrumentation amplifier

The internal structure of the IA shown in Fig. 5.4 is basically the 2-op amp topology of Section 4.1.2. However, it differs from it in two aspects: First, the gain can be changed using switches SW_1 and SW_2 and second, the addition of tunable resistors $R_{1,var} - R_{4,var}$ that are used to calibrate the IA with respect to gain mismatch and CMRR.

To determine the basic gain of the IA (i.e. neglecting the calibration), it is assumed that the following feedback resistors are used to set the gain:

$$\begin{aligned} R_1^* = R_1 + R_{1,var} = R_4 + R_{4,var} &= 7600\,\Omega \\ R_{1,2}^* = R_{1,2} = R_{3,4} &= 300\,\Omega \\ R_2^* = R_2 + R_{2,var} = R_3 + R_{3,var} &= 100\,\Omega \end{aligned}$$

In the first SW_1 and SW_2 switch positions (drawn in Fig. 5.4) the gain of the IA is given, according to (4.19) of Section 4.1.2, by $R_1^*/(R_{1,2}^* + R_2^*)$ which gives a gain of 20. Changing both the switch positions of SW_1 and SW_2 gives rise to an IA gain of 80 as the gain resistor ratio becomes $(R_1^* + R_{1,2}^*)/R_2^*$.

The M3C2 input common-mode range of the IA is determined by its gain as shown by (4.23). For the gain of 20, the maximum ICMR is given by ± 1.58 V whereas for a gain of 80 it extends to ± 1.63 V.

The input-referred noise power density $v_{ni,2IA}^2(f)$ of the IA is calculated using (4.32) which includes the thermal noise of R_1, the input-referred noise of the op amps and the 2-op amp IA gain. The verification, if the above noise specifications are met, will be illustrated for the case of the low-power, high-noise ECG specification and the low-noise, high-power EEG application.

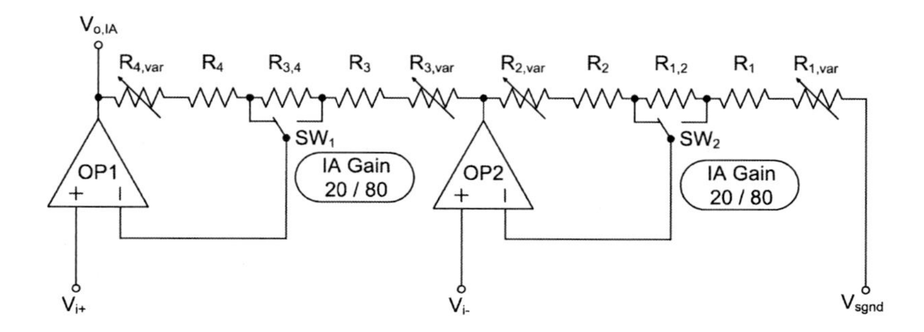

Figure 5.4: Structure of the M3C2 IA.

In the case of ECG, the IA is set to a gain of 20 and the bias current of the op amp is set to 2 μA. Using this bias setup gives an simulated op amp input-referred thermal voltage noise density of $v_{n,OP}(f) = 10.6\ \text{nV}/\sqrt{\text{Hz}}$. Considering that chopper stabilization will be used with a ratio of 1/f corner frequency f_k to chopper frequency f_{ch} of 1/25 at this bias, the noise value increases slightly to 10.63 $\text{nV}/\sqrt{\text{Hz}}$ according to (4.106). Now, determining the thermal noise of $R_1 = 7.6\ \text{k}\Omega$ and using (4.32) the M3C2 input-referred voltage noise density can be calculated to be $v_{ni,2IA}(f) = 15.44\ \text{nV}/\sqrt{\text{Hz}}$. This value gives finally a total input-referred voltage noise of approximately 1.5 μVpp for the ECG bandwidth of 0.05 – 250 Hz.

For the EEG specification the gain is set to 80 and the op amps are biased with a current of 20 μA to achieve a low-noise operation. Now, the simulated op amp input-referred thermal voltage noise density is $v_{n,OP}(f) = 3.25\ \text{nV}/\sqrt{\text{Hz}}$, $f_k/f_{ch} = 0.006$ and $R_1 = 7.6\ \text{k}\Omega$ giving rise to an input-referred IA voltage noise density of $v^2_{ni,2IA}(f) = 5\ \text{nV}/\sqrt{\text{Hz}}$. The total input-referred noise for the EEG bandwidth of 0.5 – 70 Hz becomes about 0.3 μVpp.

The values obtained show that the noise specification agree with the values in Table (5.1), the values for EP and EMG were calculated likewise and fulfill the specifications given. Finally, the static power dissipation of the IA has to be considered which is simply given by the sum of the two op amp static power dissipations. The simulated values are 686 μW and 7.2 mW for the ECG and EEG setups, respectively. These values leave enough margin for the power dissipation of the additional front-end blocks and the ADC to fit into the values given in Table (5.1).

The addition of the prototype M3C2 DC-suppression circuit to the IA (see also *Type C front-end* in Section 2.5) uses subtraction of the electrode DC-offset from the output voltage of the 2-op amp IA. For this, an additional op amp, configured as an inverting amplifier with the positive input terminal connected to an adjustable voltage $V_{DC,set}$, is placed as a second stage to the 2-op amp IA. Fig. 5.5 illustrates the functionality on the basis of a generic 2-op amp IA.

It will be now assumed that the internal 2-op amp IA output voltage $V_{o,int}$ is the sum of an amplified, time varying differential signal voltage and an amplified DC-offset voltage, i.e. $V_{o,int} = V_{sig,d} + V_{DC}$. The relation between the overall output voltage V_o and the internal 2-op amp IA output voltage $V_{o,int}$ is given by

$$\begin{aligned} \frac{V_{o,int} - V_{DC,set}}{R_x} &= \frac{V_{DC,set} - V_o}{R_y} \\ &\Leftrightarrow \\ R_y V_{sig,d} + R_x V_o &= \underbrace{(R_x + R_y) V_{DC,set} - R_y V_{DC}}_{\stackrel{!}{=} 0} \end{aligned} \tag{5.1}$$

where a simple inverting amplifier equation results when the left hand term becomes zero. This is accomplished by application of a $V_{DC,set}$ voltage of

$$V_{DC,set} = \frac{R_y V_{DC}}{R_x + R_y} \tag{5.2}$$

which results in a DC-offset free output voltage V_o.

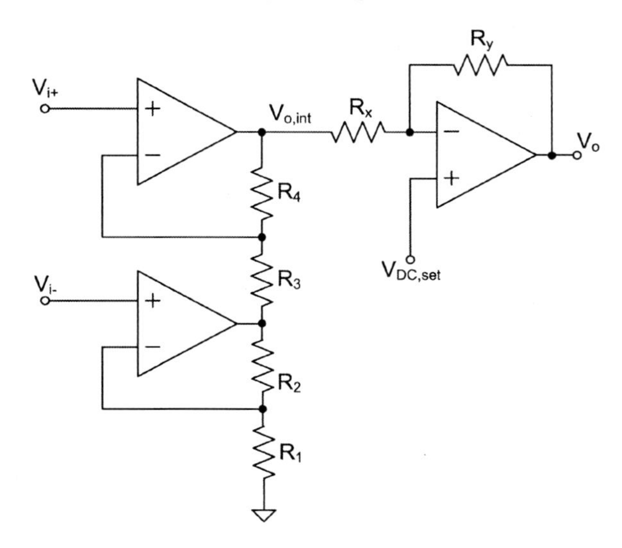

Figure 5.5: Circuit to illustrate the M3C2 DC-suppression functionality.

The DC control voltage $V_{DC,set}$ is provided by applying the digitalized signal values to a 12-bit digital-analog converter (DAC), similar to the structure depicted in Fig. 2.9 (b). The M3C3 solution uses the same DAC structure, more information on its structure are found in the description of the M3C3 in Section 5.3. It has to be noted that in order to use the DC-suppression solution the output of an additional non-inverting amplifier has to be connected to the terminal V_{sgnd} of the IA in Fig. 5.4 which can be the amplifier for the driven right leg (DRL) circuit. In this case, the amplifiers and hence their input terminals, change their function: Input terminals V_{i+} and V_{i-} in Fig. 5.4 change to $V_{DC,set}$ and V_{i+}, respectively. The input terminal V_{i-} will then be provided by the additional non-inverting amplifier. The implemented (prototype) DC-suppression serves as a test vehicle and cannot be used concurrently with the calibration circuitry which relies on the original 2-op amp IA design.

IA Calibration

Two important IA parameters are missing in the above considerations, namely the gain mismatch ΔA_{2IA} given by Eq. (4.21) and the CMRR of the IA that is mainly determined by the CMRR due to resistor mismatch, $\text{CMRR}_{\Delta R}$, which can be calculated using (4.25). Both, $A_{\Delta 2IA}$ and $\text{CMRR}_{\Delta R}$ depend on the relative error in resistor ratio of the 2-op amp IA, Δ_{2IA}, hence by changing this ratio both parameters can be set to desired values. For this, a calibration circuitry was implemented and its functionality will be shortly outlined here. Further details on this circuit can be found in [96].

The calibration circuitry is based on the tunable resistors depicted in Fig. 5.4. These resistors can be digitally controlled to have 16 different resistance values with a fixed part lying roughly in the range of $90\,\Omega$ for $R_{4,var}$ and $R_{1,var}$, and $50\,\Omega$ for $R_{3,var}$ and $R_{2,var}$. Calibration of the resistor ratio for the left hand amplifier in Fig. 5.4 is now accomplished by increasing the resistance $R_{4,var}$ while decreasing the resistance of $R_{3,var}$ and vice versa. The same arguments hold for the right hand amplifier in Fig. 5.4. In addition, the actual gain (i.e. 20 or 80) must be taken into account. A calibration sequence starts by applying signal ground to the right hand amplifier input V_{i-} thereby setting the OP2 output to signal ground. Now, a test signal (e.g. 50 Hz) is applied to V_{i+} and the IA output voltage is measured while $R_{4,var}$ and $R_{3,var}$ are changed until the specification is reached. Afterwards the CMRR calibration is performed by applying a common-mode signal to the shorted inputs V_{i+} and V_{i-} and changing $R_{1,var}$ and $R_{2,var}$ to obtain the desired CMRR value.

For finely granulated tunable resistor steps, the overall CMRR is no longer limited by the resistor mismatch, but by the open-loop gain of the operational amplifier used. The CMRR value related to the finite open-loop gain, $\text{CMRR}_{\Delta OP}$, determined by simulation of the OP350a, is 88 dB and 107 dB for the setups using the bias currents $I_{bias} = 2\,\mu$A and $I_{bias} = 20\,\mu$A, respectively. The increase of the $\text{CMRR}_{\Delta OP}$ for higher bias current is due to the higher bandwidth of the op amps.

5.2.3 ADC

The analog-to-digital conversion of each channel is obtained using a $\Sigma\Delta$-ADC. The ADC of each channel (see also Fig. 5.2) consists of a 2nd-order$\Sigma\Delta$-modulator with differential input, which samples the input signal with an oversampling ratio of 256. This stage is followed by a decimation filter to digitally low-pass filter the signal and reduce the data rate by 256; the decimation filters exhibit a 20-bit output bus. The modulator is based on a time-sharing single amplifier structure [97] and the decimation filter uses polyphase decomposition of a 3rd-order comb filter [98]. The ADC is programmable with respect to the output data rate which can be varied between $1 - 50$ kHz. Its effective number of bits (ENOB) depends on the data rate applied with the lowest accuracy being 14 bits. The power dissipation changes also with the applied data rate and lies between $240\,\mu\text{W} - 3$ mW. Adding these values to the power consumption of the analog front-end results in a channel power dissipation that still complies with the specifications given in Table 5.1.

5.2.4 Respiration Circuitry

A novel feature regarding integrated biomedical signal acquisition systems is the addition of a respiration measurement unit (RU) to be activated during ECG measurement. The system is fully integrated on-chip (except for an external capacitor) and uses the channels of the IA as part of its signal chain. The respiration monitoring principle, based on bioelectric impedance measurement (BEI) [14] on the patients' thorax, is shown on the left hand side of Fig. 5.6. Applying a differential test signal in the kilohertz range (the signal generator has an output impedance of $2Z_{sig}$) to the body-electrodes and the thorax impedance, Z_{Body}, forms voltage divider. The action of patient breathing changes the thorax impedance to $Z_{Body} = Z_{Body,0} + \Delta Z_{Body}$, with $Z_{Body,0}$ being the mean overall thorax impedance and ΔZ_{Body} denotes deviation due to respiration. The mean body impedance is mostly resistive and lies in the range of $\approx 300\,\Omega$ with ΔZ_{Body} being $1 - 2\,\Omega$ per liter of lung volume change (the volume change at rest for an adult is about 0.4 L) [14]. As a result, the voltage $V_{CRin,diff}$ consists of the carrier signal modulated by the change of body impedance. The maximum current flowing through the thorax has to be limited in order to not exceed medical device regulation. A major advantage of this solution is its incorporation into two ECG channels which allows to concurrently measure the respiration and ECG signals, i.e. no additional respiration electrodes are needed.

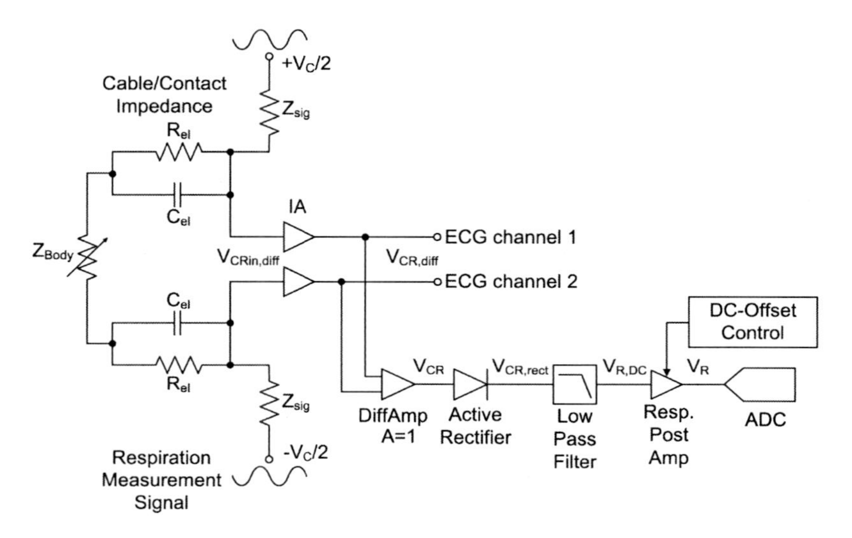

Figure 5.6: Respiration measurement circuit.

In the present system, the on-chip oscillator is used to generate a differential sine-signal having a frequency of 40 kHz. The modulated carrier signal $V_{CRin,diff}$ is then amplified by the IA and the differential signal $V_{CR,diff}$ is tapped and applied to the on-chip respiration module as depicted in Fig. 5.6. The low-pass filter of the regular channel located before the AD-conversion removes the 40 kHz signal, so that no oscillator signal is present in the ECG signal. Within the respiration block the signal is first converted from differential to single-ended using an op amp configured as a differential amplifier with unity gain. Next, the signal is applied to an active full-wave rectifier and a subsequent Sallen-Key low-pass filter to extract the very low frequent (≤ 1 Hz) respiration signal.

The resulting signal $V_{R,DC}$ with both, the respiration signal and the rectified sine wave contains a large DC value that equals $2V_C/\pi$ for a pure sine wave. A direct amplification of the $V_{R,DC}$ voltage containing the very small respiration signal would therefore result in a saturation of the amplifier. The use of a simple high-pass filter was impeded by the very low frequent nature of the respiration signal and very slow DC-offset drifts induced, for example, by sweating of the patient. This would give rise to impractical high device sizes. Therefore, another way to eliminate the slowly varying DC-offset had to be followed.

A solution to this problem was found by incorporating a DC-offset control within the postamplifier, the solution is shown in Fig. 5.7. With this system being a prototype, gain programmability was added to the postamplifier to determine an optimal amplification. As a result, the postamplifier can be set to have a gain of 25, 50, 100 or 125.

The postamp is composed of three non-inverting amplifiers (labeled by I, II and III in Fig. 5.7a) with the reference voltage of amplifier I provided by the output of amplifier II. Amplifier II has signal ground connected to its input terminal and its output voltage is controlled digitally by changing the tap point within R_{12b}/R_{R21b} consisting of 32 resistors and access switches, see Fig. 5.7b. Amplifier III is fed by the output voltage of amplifier I, $V_{o,a}$, to achieve the overall gain. The resulting voltage V_R is monitored by two comparators comparing the voltage V_R to the reference voltages V_{ref+} and V_{ref-}. If V_R exceeds either the upper or lower reference voltage, a digital signal (*over* or *under*) is generated by one of the comparators. This signal is used by a digital control block using 32 output lines to adjust the tap point of R_{12b}/R_{R21b} to a value which reduces the DC-offset causing the amplifier saturation. The overall gain setting is controlled by a 4 bit gain control line setting the tap point within R_{1c}/R_{R2c}, see also Fig. 5.7c.

The signal V_R is finally digitalized using an 8-bit successive approximation ADC that was available as an IP-block. All op amps used in this circuit are a modified version of OP350a of Section 4.3. The op amps have smaller Miller capacitances and input transistors in comparison to the OP350a design, and hence a reduced area consumption. The respiration measurement unit can also be used to detect open-leads by measuring the absolute impedance between leads during an ECG recording. The power consumption of the RU is about 2.6 mW using a sampling frequency of 200 Hz.

5.2.5 Digital Circuits including DSP

The M3C2 digital circuit is composed of a timing and control unit (TCU), a programmable clock unit (PCU), the I/O-interface and the on-chip digital signal processor (DSP). The TCU acts as an interface between each channel and the DSP. On the one hand it stores the digitalized biomedical signal values which are accessed by the DSP, and on the other hand the DSP sets registers within the TCU that control the setting of the analog front-end like gain or bandwidth. The PCU is used to generate the internal clock signals derived from an external main clock. The main I/O interface of the system is composed of three UARTs and a 16-bit parallel port.

The on-chip DSP (model C32025TX - provided as a Verilog IP core by Evatronix SA [99]) was implemented in silicon by Dipl.-Ing. Wjatscheslaw Galjan. It

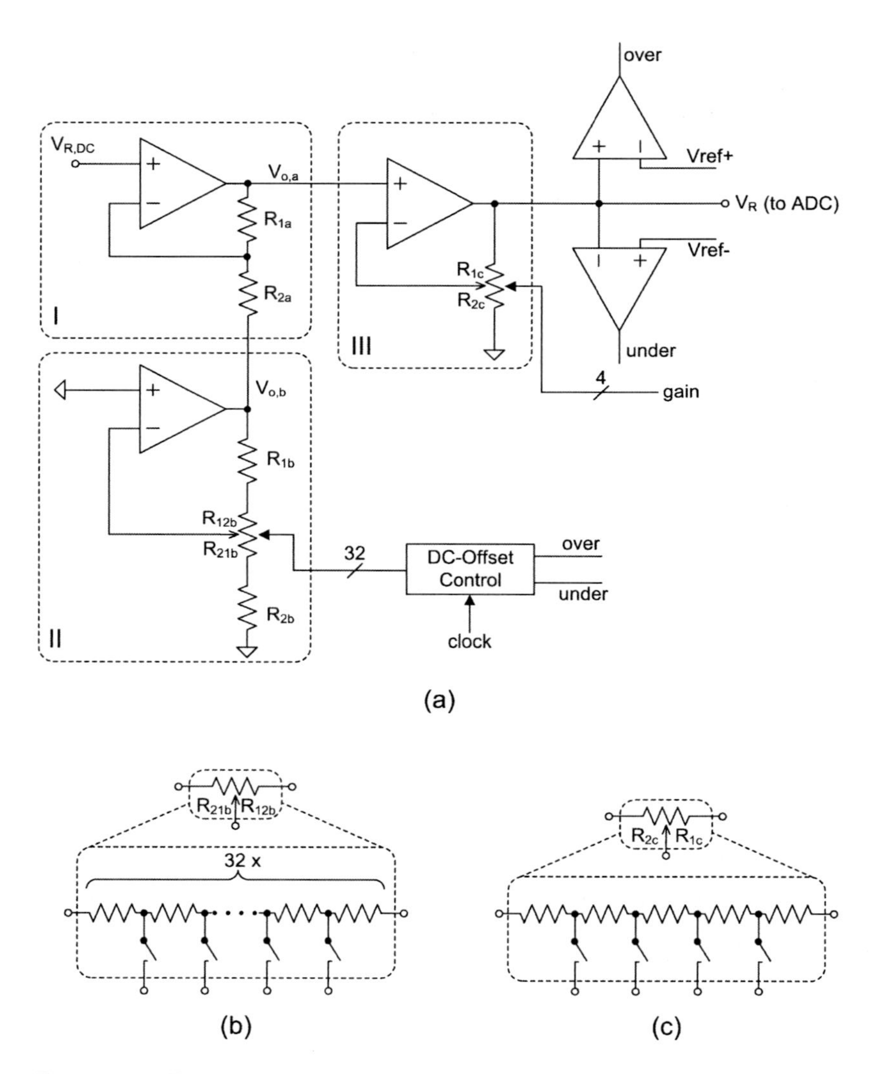

Figure 5.7: RU postamplifier with offset control circuit. The main circuit is shown in (a) and the implementations of the feedback resistors for amplifier II and III are shown in (b) and (c), respectively.

is a 16-bit fixed point DSP which is instruction compatible to the Texas Instruments TMS320C25. In addition to the internal memory, two 4k words SRAMS were integrated on-chip as program and data memory, respectively. The DSP is used as the main system control unit and can be used to implement various signal processing tasks like digital filtering, biomedical signal feature extraction and signal compression, an example application is given in [100].

5.2.6 Autocalibration signal and BIST

The on-chip analog oscillator is used to produce precise single-tone test signals for use in the autocalibration of the front-end [96]. Additionally, the signal generator provides the test signal for the respiration measurement. The single-ended to differential conversion needed for the respiration block is achieved by applying the output signal of the oscillator to an inverting amplifier also implemented in the system. The architecture of the signal generator is based on oversampling $\Sigma\Delta$-modulation [101] and provides programming capability with respect to signal amplitude and frequency. Thus, the implemented version allows programmability of the test signal frequency between 50 Hz $-$ 40 kHz and the peak-to-peak output amplitude range is spanning from 25 mV to 1.5 V.

The implemented autocalibration and built-in self test (BIST) routines [96] are used to calibrate each front-end of the channels separately with respect to gain and CMRR and additionally test for the input-referred noise by connecting the inputs of the IA to signal ground. For this, a low-noise input selector circuit was placed in front of the IA and allows to connect the appropriate IA input terminal to the signal input, the test signal, signal ground or the DC-suppression voltage. After powering the M3C2 up, a test signal is used to calibrate the gain, followed by a CMRR calibration step. Afterwards, the input terminals are connected to signal ground to determine the input-referred noise. The calibration and BIST system could be used, in general, to test additional parameters like inter-channel crosstalk or ADC accuracy.

5.2.7 Realization

The M3C2 SoC was realized in a 3.3 V, 350 nm CMOS technology and has an area of 6800 μm $\times$ 5500 μm, a micrograph of the chip is shown in Fig. 5.8. The M3C2 chip has a pin count of 144 and is packaged in a PGA (Pin Grid Array) package for testing in a dedicated measurement setup.

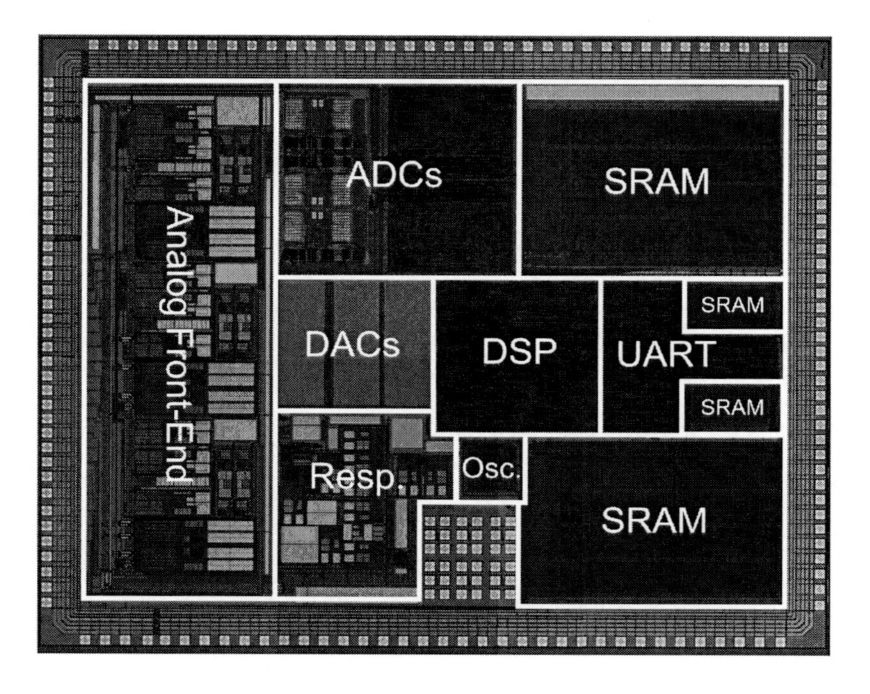

Figure 5.8: M3C2 SoC chip photograph.

5.2.8 Results

Power Dissipation

The power dissipation of the analog front-end excluding the ADCs was measured for the lowest and highest AFE power setting $I_{bias,low} = 2\,\mu$A and $I_{bias,high} = 20\,\mu$A, respectively. At lowest bias setting, a single AFE channel has a power dissipation of 4 mW and for the highest bias current 10 mW are dissipated. The power consumption of a single ADC was separately measured and amounts to 0.24 mW and 3 mW for a sampling frequency of 1 kHz and 50 kHz, respectively. Considering the design specifications given in Table 5.1, these results show that even at the highest sampling rate of 50 kHz and using $I_{bias,low}$ for ECG or $I_{bias,high}$ for EEG, the power dissipation fulfills the conditions given.

The main power dissipation of the digital circuit is due to the DSP and the three UARTs. The implemented DSP consumes 0.62 mW/MHz being in idle mode and 1.77 mW/MHz running at full load. The DSP was tested for its maximum frequency of operation which is 60 MHz. The UARTs consume a noticeable power of 1.47 mW/MHz only at full speed.

Noise Performance

The total input-referred noise was measured for the lowest and highest bias current setup. Table 5.2 shows the results including the design specification and calculated values.

Bandwidth (Signal type)	Condition	Value	Specified	Calculated
0.05 Hz - 250 Hz (ECG)	$I_{bias,low}$	1.47 μVpp	$<$ 5 μVpp	1.5 μVpp
	$I_{bias,high}$	0.96 μVpp		
0.5 Hz - 70 Hz (EEG)	$I_{bias,low}$	0.89 μVpp	$<$ 1 μVpp	
	$I_{bias,high}$	0.61 μVpp		0.3μVpp
0.1 Hz - 3 kHz (EP)	$I_{bias,low}$	3.55 μVpp	$<$ 2.2 μVpp	
	$I_{bias,high}$	1.78 μVpp		
0.01 Hz - 5 kHz (EMG)	$I_{bias,low}$	4.32 μVpp	$<$ 6 μVpp	
	$I_{bias,high}$	2.08 μVpp		

Table 5.2: Total input-referred voltage noise of the M3C2 (chopper stabilization enabled). The calculated values are from Section 5.2.2.

The ECG, EEG and EMG design specifications can be fulfilled using the lowest and highest bias setups. To comply with the EP bandwidth specification, the AFE has to be biased by $I_{bias,high}$. The calculated value for the ECG bandwidth matches closely the measured values whereas the noise for the EEG is approximately two times higher. This value is attributed to additional noise from the front-end not considered in the derivation of the IA noise. The NEF of the complete front-end can be calculated to be 21.2 and 26.3 for the ECG and EEG setups, respectively. The NEF of the IA can only be approximated by using the calculated power dissipation values and measured noise values which present an upper limit to the actual IA noise behavior. This calculation results in an IA NEF of 8.8 for ECG and 22.3 for EEG applications. The values match approximately the NEF trend line (NEF $= 10 \ldots 20$) for present-day biomedical amplifiers [91].

Respiration Unit

To quantify the performance of the respiration unit, an accurately defined test setup to simulate the thorax and cable impedances was assembled. The setup uses an variable resistor of 300 Ω as a replacement for Z_{Body} and 1 kΩ resistors to approximate the cable impedance at the test signal frequency of 40 kHz. By slowly changing the variable resistor within the range of $1\,\Omega$ an RU output voltage change of 12.5 mV was measured.

The DC-offset control was tested by applying a 40 kHz test signal with a slowly increasing amplitude. The action of the rectification of the sine wave with increasing amplitude results in a slowly increasing DC-offset. The gain of the postamp was set to 100 and the RU output voltage recorded is shown in Fig. 5.9. When the DC-offset reaches values near to the positive supply rail the DC-offset control changes the tap point of the R_{12b}/R_{R21b} resistor ratio of amplifier II depicted in Fig. 5.7, thereby restoring a lower output DC level (The ripple on the test signal is an artifact from the measurement setup). Likewise, the amplitude of the test signal was swept from a high to a low value which resulted in an output voltage that resembles Fig. 5.9 with the time line reversed.

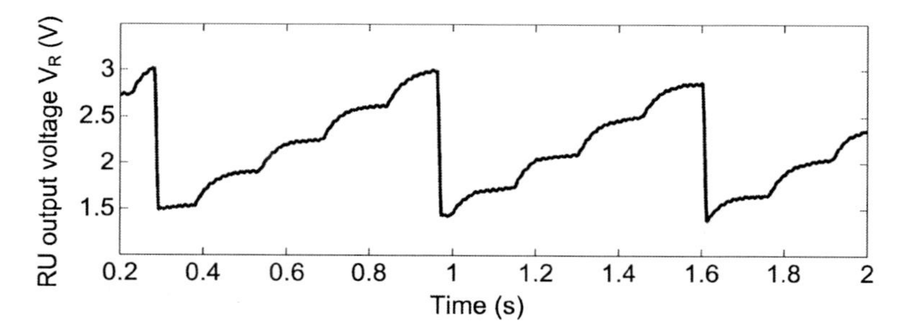

Figure 5.9: RU DC-offset control preventing the output voltage from saturation.

In order to test the circuit under realistic conditions, a RU recording was performed with a test person (30 years old, male, 70 kg) using two disposable, self-adhesive Ag/AgCl electrodes placed across the chest. Fig. 5.10 shows the absolute voltage change of V_R referenced to a point at approximately half of the peak-to-peak value of the respiration signal.

Autocalibration and BIST

The autocalibration and BIST circuits were used to calibrate the AFE with respect to gain accuracy and CMRR. The measurements were conducted using the 2-op amp IA configuration, i.e. without using the DC-suppression setup. The results of a single channel calibration using a 230 Hz test signal are listed in Table 5.3. For each setup, at least one value meets the specified values. The use of 230 Hz for measuring the CMRR poses no problem as the results do not differ significantly from the CMRR at 50 Hz when considering resistor mismatch only. However, the

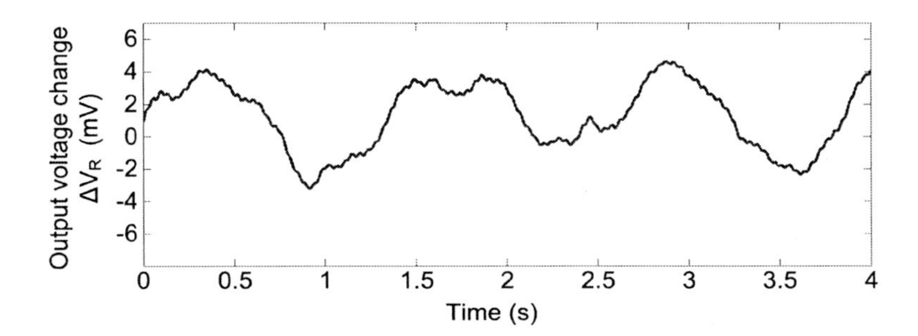

Figure 5.10: Respiration signal of a test person using the RU and two Ag/AgCl electrodes.

CMRR limited by the open-loop gain of the op amp, $\mathrm{CMRR}_{\Delta OP}$, decreases for higher frequency. Thus, the use of 230 Hz would even underestimate the actual CMRR at 50 Hz if it is dominated by $CMRR_{\Delta OP}$. More details on the autocalibration/BIST and signal generator results are given in [96].

A_{2IA}	Number of matching codes		
	$\Delta A_{2IA} < 1\%$	CMRR > 63 dB	Peak CMRR value
20	3	4	≈ 80 dB
80	1	7	≈ 80 dB

Table 5.3: Number of tunable resistor code values (out of 16) matching the specifications of Table 5.1 and the peak CMRR value measured.

5.2.9 Results overview and Application Examples

The main M3C2 SoC characteristics have been compiled into Table 5.4. Two application examples are given below to verify and demonstrate the capabilities of the system.

24h x 7d portable ECG System

In order to demonstrate the ability of the M3C2 SoC in a portable, long-term ECG recording device, a prototype system has been developed [100]. The system is made up of a 60 mm × 100 mm printed circuit board (PCB) containing the M3C2 chip, a Compact Flash (CF) card interface and additional support devices. A photograph of the PCB is shown in Fig. 5.11a. The system can be operated for up to 10 days using

Parameter	Condition/Remarks	Value
Supply Voltage		3.3 V
Number of channels		3
Gain Settings		20/80/320
Bandwidths	programmable	2.4/12.5 kHz
Power Dissipation		
AFE	programming range	$4 - 10$ mW per ch.
ADC	$1 - 50$ kHz sampling rate	$0.24 - 3$ mW
DSP	idle to full load	$0.62 - 1.77$ mW/MHz
Total input-referred noise[1]	0.05 Hz - 250 Hz (ECG)[2]	$\leq 1.47\ \mu$Vpp
	0.5 Hz - 70 Hz (EEG)[3]	$\leq 0.61\ \mu$Vpp
	0.1 Hz - 3 kHz (EP)[3]	$\leq 1.78\ \mu$Vpp
	0.01 Hz - 5 kHz (EMG)[3]	$\leq 2.08\ \mu$Vpp
Max. CMRR	@ 50/60 Hz	80 dB
Gain error		$< 1\%$
Die area		37.4 mm^2
Pin number		144
Package		PGA

[1] Vpp is calculated as the 6-fold value of the root-mean-square voltage noise Vrms
[2] Low-power mode
[3] High-power mode

Table 5.4: Main characteristics of the M3C2 SoC.

two standard AA batteries. A 50 Hz notch filter and a 100 Hz low-pass filter were implemented using the on-chip DSP for signal conditioning. The ECG recording is stored on the CF card and can be read out by a personal computer. For verification, a 3 channel ECG recording was taken from a test person (28 years old, male, 72 kg), the results are shown in Fig. 5.11b.

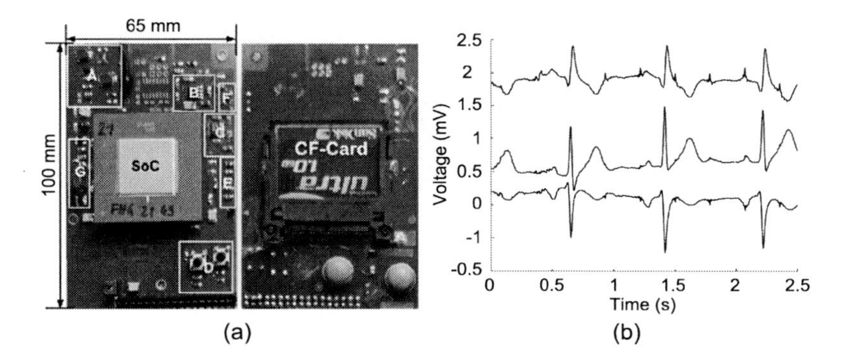

Figure 5.11: Portable 24h x 7d ECG system using the M3C2. (a) Top and bottom view of the system PCB; (b) Three channel ECG recording.

Mobile EEG System

An experimental mobile EEG signal acquisition system using the M3C2 was developed for testing of different algorithms to suppress noise and artifacts of a multi-electrode recording. The setup uses three EEG electrodes located closely to each other at the occipital region. A recorded EEG alpha-wave using this setup has been shown in Chapter 2, Fig 2.3. The realized system PCB has the dimensions of 60 mm $\times$ 75 mm. Fig. 5.12a shows a photograph of the PCB and in Fig. 5.12b a photograph of an EEG measurement session is shown. More details of this system, in particular the algorithms used, can be found in [12].

5.3 M3C3 System Solution

5.3.1 Introduction and System overview

The M3C3 is a programmable, 10 channel biomedical signal acquisition chip including a DC-suppression circuit and digital I/O for data output and configuration [102].

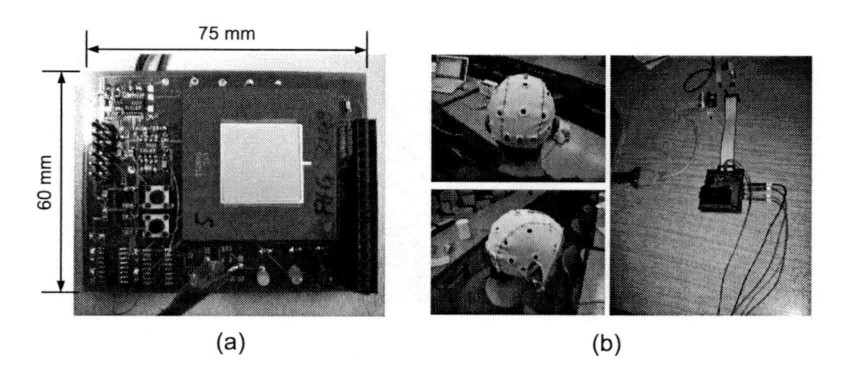

Figure 5.12: Mobile EEG System using the M3C2. (a) System PCB photograph; (b) EEG recording session.

The design was jointly carried out by the author, Dipl.-Ing. Wjatscheslaw Galjan with a main contribution to the topology of the IA and Dipl.-Ing. Kristian Hafkemeyer with a focus on the CMRR calibration layout and according simulation. The ADC and the digital part were derived from a previous development by Dipl.-Ing. Sven Vogel.

The main design target was to create a system for use in a high electrode count (100 or more) EEG system using multiple M3C3 in parallel. Therefore, the system does not include a DSP, however its front-end resembles the AFE of the M3C2 with the addition of a different DC-suppression scheme. Like the M3C2, the system can be used for low-power ECG recording. An overview of the M3C3 architecture is depicted in Fig. 5.13.

The system channels are composed of an analog front-end with differential output and a $\Sigma\Delta$-ADC formed by the $\Sigma\Delta$-modulator of each channel and a decimation filter. The front-end exhibits an additional input terminal for DC-offset suppression. This input voltage is generated (for each channel) by its associated DAC with values set by a digital control circuit. Additional digital circuits include the timing and control of the system, clock setting and distributions, registers for setup information and the digital I/O. The I/O interface includes a high speed serial dataport for the output of measurement data and sample numbering information (frame counter). A bidirectional serial interface is used for configuration purposes.

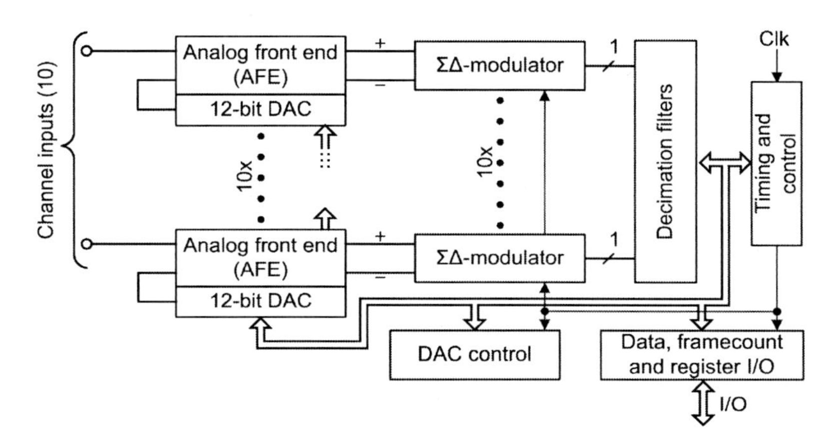

Figure 5.13: Block diagram of the system-on-chip.

5.3.2 Analog Front-End

In order to extend the flexibility of the front-end with regard to the removal of electrode inferred DC voltages both, the type A and C (see Section 2.5) suppression schemes were employed in the M3C3. These can be used individually or combined, where a joint action of both schemes gives rise to the highest possible input-referred DC-offset range for the M3C3 system. The internal structure of the M3C3 analog front-end is shown in Fig. 5.14.

The instrumentation amplifier input includes the signal terminals V_{i+}, V_{i-}, a signal ground connection V_{sgnd} and an additional DC-suppression input $V_{DC,set}$, normally not found in standard IAs. The gain of the IA was set to a constant value of $A_{IA} = 4$ and the CMRR of the instrumentation amplifier is calibrated with respect to resistor mismatch. The output voltage of the M3C3 IA, $V_{o,IA}$, is fed off-chip for connection to an optional external high pass filter. This filter can be realized by a first order RC high-pass with its RC time constant set to the lowest signal frequency. The PGA input $V_{i,PGA}$ is short-circuited off-chip to the IA output when no external filter is used, otherwise it connects to the output of the external high-pass filter. An additional feature regarding the external high-pass filter is a chip-internal switch that connects the input of the PGA via the small resistor R_{Block} to signal ground. The switch can be activated to quickly discharge the external capacitor of the high-pass filter if events like external stimulation or electrode movement induce a large shift in the DC-offset. Otherwise, a relative long time constant for restoring the system back to equilibrium would result due to the large time constant of the

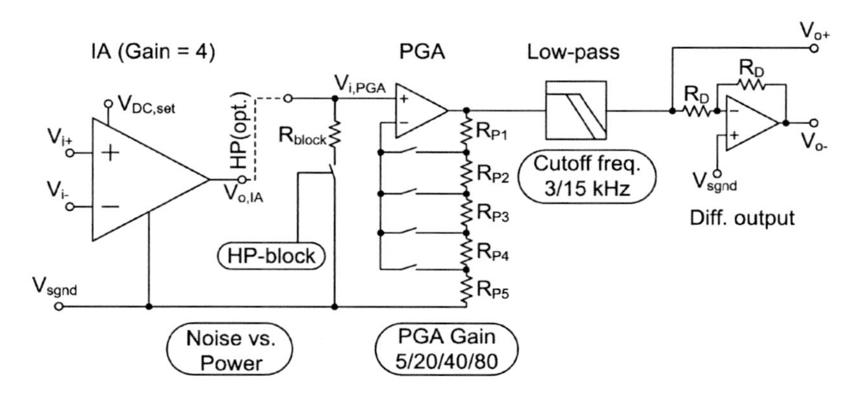

Figure 5.14: M3C3 analog front-end with programmable parameters in rounded rectangles.

external filter.

The PGA follows the design approach in Section 4.1.3 using four programmable gain factors, namely 5, 20, 40 and 80. The values of the five gain setting resistors are given in Table 5.5. Taking the constant IA gain of 4 into account results in the overall channel gains of 20, 80, 160 and 320. The PGA is followed by an anti-aliasing filter implemented in a Sallen-Key topology (see Fig. 4.10 in Section 4.1.4). The corner frequency of the low-pass filter f_c is programmable and can be set to 3 or 15 kHz, the resistor and capacitor values used for each setup are listed in Table 5.5.

The single-ended to differential conversion of the signal needed for the differential input of the $\Sigma\Delta$-modulator is accomplished using the output voltage of the low-pass filter directly (V_{o+}) and inverting V_{o+} with the help of an op amp connected as an unity-gain, inverting amplifier. The size of its feedback resistors is also presented in Table 5.5.

For the IA and the PGA, the OP350a design of Section 4.3 was used, whereas for the low-pass filter and inverting amplifier a modified version of the OP350a was taken. This version exhibits a smaller layout area, but lower noise performance, hence it is used only in the last two front-end stages. Table 5.6 gives an overview of the M3C3 analog front-end design specifications.

Front-end block	Comment	Resistor/Capacitor values
PGA		$R_{P1} = 30.4$ kΩ, $R_{P2} = 5.7$ kΩ, $R_{P3} = 950$ Ω
		$R_{P4} = 475$ kΩ, $R_{P5} = 475$ kΩ
Sallen-Key LP[1]	$f_c = 3$ kHz	$R_1 = 952.5$ kΩ, $R_2 = 952.5$ kΩ
		$C_1 = 38\,\mu$F, $C_2 = 76\,\mu$F
	$f_c = 15$ kHz	$R_1 = 190.5$ kΩ, $R_2 = 190.5$ kΩ
		$C_1 = 38$ pF, $C_2 = 76$ pF
Inv. Amplifier		$R_D = 10$ kΩ

[1] See circuit schematic in Fig. 4.10. Switches within the circuit are used to change the effective resistor values.

Table 5.5: Resistor and capacitor values (rounded) for all front-end circuits except the IA.

Parameter	Condition/Remarks	Value
Signal input range	Ext. programmable	± 5 mV
		± 20 mV
		± 40 mV
		± 80 mV
Total input-referred noise[1]	0.05 Hz - 250 Hz (ECG)	< 5 μVpp
	0.05 Hz - 70 Hz (EEG)	< 1.0 μVpp
	0.01 Hz - 5 kHz (EMG)	< 6.0 μVpp
	0.1 Hz - 3 kHz (EP)	< 3 μVpp
Power dissipation	ECG	7 mW
(incl. AFE and ADC)	EEG	25 mW
CMRR	@ 50 Hz	> 80 dB
Offset comp. range	DC-suppression.	± 600 mV
	DC-suppr. & HP	± 1000 mV

[1] The peak-to-peak voltage noise Vpp is calculated as the 6-fold value of the root-mean-square voltage noise Vrms

Table 5.6: M3C3 analog front-end design specifications.

Instrumentation Amplifier with DC-suppression and CMRR Calibration

The M3C3 instrumentation amplifier depicted in Fig. 5.15 is based on a standard 2-op amp IA topology with the addition of a DC-suppression and a CMRR calibration. The standard 2-op amp IA structure is therefore extended using a third op amp and a programmable calibration resistor divided in $n = 64$ step sizes of ΔR [103]. The functionality of the DC-suppression can be explained best for the static case by considering the internal signal $V^*_{o,OP1}$ in Fig. 5.15 as the control voltage to remove a DC-component from the static input signal $(V_{i+} - V_{i-})$.

The voltage $V^*_{o,OP1}$ is connected via R_{3b} to the inverting input of OP1. Hence, a positive (or negative) voltage drop of $(V^*_{o,OP1} - V_{i+})$ across R_{3b} develops a DC current I_{R3b} flowing to (from) the node connecting R_{3a}, R_{3b} and R_4. For a regular 2-op amp IA, no current flows at the inverting OP1 input and I_{R3a} (I_{R3}) equals $(V_{i+} - V_{o,OP2})/R_{3a}$. If a DC-offset is present at the input voltage $(V_{i+} - V_{i-})$, $V^*_{o,OP1}$ is set to a such value (see equation (5.3) below) that the DC current I_{R3b}, which is forced to R_4, gives rise to the desired DC-shift of the output voltage of the IA.

However, in the dynamic case I_{R3b} adds an additional AC current to the resistor R_4 which changes the overall 2-op amp IA transfer characteristic. Therefore, the ratio of the feedback resistors R_1-R_4 has to be set to a specific value in order to restore the default 2-op amp IA operation.

A detailed analysis of the IA output voltage with respect to the IA input voltages is given in [103]. It includes the calibration resistor setting $n \cdot \Delta R$ and any parasitic resistor R_{par} from signal ground to R_4. The IA output voltage is accordingly given by

$$V_{o,IA} = V_{i+}\left(1 + \frac{R_4}{R^*_{3b}} + \frac{R_4}{R_{3a}}\right) - V_{i-}\frac{R_4}{R_{3a}}\left(1 + \frac{R_2}{R^*_1}\right) - V_{DC,set}\frac{R_4}{R^*_{3b}} \quad (5.3)$$

with $R^*_{3b} = R_{3b} + n \cdot \Delta R$ and $R^*_1 = R_1 + R_{par}$.

The DC-suppression of the IA is expressed by the term $-V_{DC,set}(R_4/R^*_{3b})$ in (5.3). In case of differential input voltages, a differential gain with respect to $V_{i+} - V_{i-}$ is only given, if both voltages are amplified by the same magnitude, i.e. if the condition

$$1 + \frac{R_4}{R^*_{3b}} + \frac{R_4}{R_{3a}} = \frac{R_4}{R_{3a}}\left(1 + \frac{R_2}{R^*_1}\right) \quad (5.4)$$

is satisfied. In this case the gain is given by

$$A_{IA} = 1 + 2\frac{R_4}{R_3} \quad (5.5)$$

using $R_3 = R^*_{3b} = R_{3a}$ and neglecting R_{par}. The maximal DC-offset, that can be suppressed by this circuit, is limited by the maximal or minimal $V_{DC,set}$ voltage that can be applied, i.e. $V_{DD}/2$. When considering the output of the IA and assuming

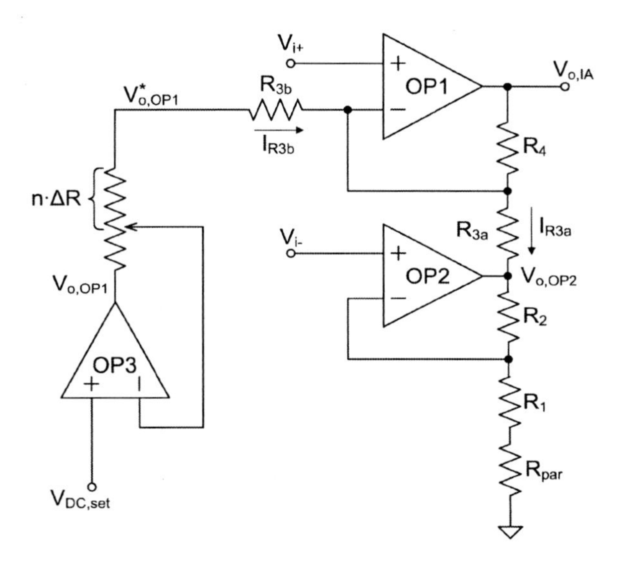

Figure 5.15: M3C3 instrumentation amplifier incorporating DC-suppression and CMRR calibration.

$R_3 = R^*_{3b} = R_{3a}$, this voltage is scaled by R_4/R_3 according to (5.3). Dividing this value by the gain of the IA gives the range of DC-offset suppression:

$$V_{i,DC} < \pm \frac{V_{DD} R_4}{2 A_{IA} R_3} \tag{5.6}$$

The gain of 4 in the present design was achieved using $R_1 = 0.6$ kΩ, $R_2 = 1$ kΩ, $R_3 = 1$ kΩ and $R_4 = 1.5$ kΩ. These resistor values and the supply voltage of V_{DD} give rise to a maximum input-referred DC-offset voltage of ±619 mV. If the external high-pass filter is additionally used for a PGA gain of 80, the suppression range increases to ±1000 mV.

The CMRR calibration of the IA is based on changing the tap point in the high-resistive feedback path of OP3 to reduce the resistor mismatch and hence increase $\text{CMRR}_{\Delta R}$. The number n of resistor step sizes ΔR changes the value of $R^*_{3b} = R_{3b} + n \cdot \Delta R$ and hence the resistor ratio of R_4 to R_{3a} and R_{3b} whereas the ratio $R_2/(R_1 + R_{par})$ stays fixed. The resulting $\text{CMRR}_{\Delta R}$ has been analyzed in [103] and can be calculated to be

$$\text{CMRR}_{\Delta R} = \frac{\frac{1}{R_4} + \frac{1}{R_{3b}^*} + \frac{2}{R_{3a}} + \frac{R_2}{R_1 R_{3a}}}{\frac{1}{R_4} + \frac{1}{R_{3b}^*} - \frac{R_2}{R_1 R_{3a}}} \tag{5.7}$$

In order to achieve a $\text{CMRR}_{\Delta R} > 80$ dB, $n = 64$ step sizes with a value of $\Delta < 0.26\,\Omega$ were used.

The M3C3 IA input-referred noise PSD can be estimated roughly by neglecting the noise of the feedback resistors and assuming that chopper stabilization is enabled. As already described in Section 5.2.2, the OP350a exhibits an input-referred voltage noise density of 10.6 nV/$\sqrt{\text{Hz}}$ and 3.25 nV/$\sqrt{\text{Hz}}$ for the lowest and highest bias current I_{bias}, respectively. The output noise PSD of the IA is calculated using the transfer function (5.3) with $R_3 = R_{3b}^* = R_{3a}$, neglecting R_{par}, and considering the input-referred noise of each op amp, i.e.

$$v_{no,IA}^2(f) = v_{n,OP}^2(f)\left(1 + \frac{R_4}{R_3} + \frac{R_4}{R_3}\right) + v_{n,OP}^2(f)\frac{R_4}{R_3}\left(1 + \frac{R_2}{R_1}\right) + v_{n,OP}^2(f)\frac{R_4}{R_3} \tag{5.8}$$

The corresponding input-referred noise PSD can be used to estimate the noise performance of the IA. For the ECG bandwidth and lowest power setting, a calculation results in a total input-referred noise of 1.48 μVpp. In case of an EEG bandwidth and highest bias current, the noise reduces to 0.24 μVpp. However, at such a low-noise level the influence of the feedback resistors can no longer be neglected. The feedback resistor sizes given above were determined by simulation to contribute about 20% to the noise of the IA. The noise for the EEG bandwidth increases therefore to approximately 0.31 μVpp. In addition, the noise of the subsequent front-end stages has to be added at the lowest noise setup. The resulting input-referred AFE noise for the EEG bandwith and highest bias current setup was simulated as 0.44 μVpp. The noise performance for the EP and EMG bandwidths can be likewise calculated.

The static power dissipation of the IA is given by three times the static power dissipation of a single OP350a. For the lowest and highest bias current, this results in an IA power dissipation of 1 mW and 10.8 mW, respectively, leaving enough "*headroom*" for the power dissipation of the remaining front-end blocks and the ADC.

With the resistor values used as given above, the input common-mode range of the M3C3 IA (with $V_{i+} = V_{i-}$ and $V_{DC,set} = 0$) is limited by the bottom amplifier OP2 (see Section 4.1.2) and results in an ICMR of ± 1.1 V.

5.3.3 ADC

Similar to the M3C2, the ADC is made up of a 2nd-order $\Sigma\Delta$-modulator with differential output and a decimation filter. The 20-bit output port of the decimation filter delivers the values of the digitized input voltage with a data rate which depends on the programmable oversampling ratio (OSR) and main clock division by an integer n. The OSR is programmable between 256 ± 128 and n can be set to values of 0 to 249. For example, using a main clock of 12.8 MHz and OSR=256, the data rate varies between 200 Hz and 50 kHz.

The topology of the discrete-time 2nd-order $\Sigma\Delta$-modulator using two OTAs is derived from [104] using capacitor values for the switched-capacitor integrators as given in [25]. The decimation filter implements a 3rd-order low-pass filter and decimates the data to the desired rate.

5.3.4 DC-suppression Circuit

When the DC-suppression circuit is used, each of the 10 on-chip DACs associated to one of the AFE channels generate an analog voltage that is routed to an output pin of the M3C3 chip. This pin is connected to the IA input pin $V_{DC,set}$ and an external bypass capacitor in the range of $1\,\mu$F to $10\,\mu$F. The capacitor is used to smooth the steps in the output voltage of the DAC due to changing values and to reduce the noise of the DAC output resistance of 100 kΩ

The 12-bit DACs were implemented using a R-2R architecture [105]. A digital integrator was additionally developed and is included tenfold in the M3C3 to implement a suppression scheme as outlined in Section 2.4.2 for each channel. The digital input values of the DACs are stored in a configuration register that can be written in two ways: either by using the on-chip digital integrator or from outside the M3C3 via the serial control interface. The second way of setting the DAC output voltage has been added to increase flexibility. Each of the 12-bit DACs dissipates a power of about 100 μW.

5.3.5 Digital Circuit and I/O

The digital circuits of the M3C3 include a timing and control block to control the ADCs, generate the chopper stabilization clocks, and for programming purposes, e.g. to program the AFE setup. The implemented high speed data transfer protocol allows to daisy-chain up to 16 M3C3s using a 4-bit chip address.

The serial configuration port realizes a flexible interface for setting up the M3C3 externally. Its main features are:

- Bidirectional data transfer
- Maximum bus speed: 100 kHz
- 4-bit chip address input
- Access to one of 100 on-chip 8-bit registers
- Control of 16 general purpose outputs (GPOs) via the on-chip registers

The channel data output of the M3C3 is realized by a high speed serial dataport running at main clock frequency of the system. The serial output data format is organized into 22 × 16-bit packets including:

- Chip address
- For each of the 10 channels:
  - Output data
  - Frame counter value
  - Chip address
  - DAC value

5.3.6 Implementation

The realized M3C3 system, shown in Fig. 5.16, was implemented in a 3.3 V, 350 nm CMOS technology. It occupies an area of 8000 μm × 5880 μm and has a pin count of 144. For the M3C3 packaging, a PGA package type as well as a BGA (ball grid array) were used. The BGA package minimizes the area of the M3C3 needed on a printed circuit board to 13 mm × 13 mm

5.3.7 Measurements

Analog Front-end

The AFE was tested for the overall gains of 20, 80, 160 and 320 and low-pass filter cut-off frequencies of 3 kHz and 15 kHz. Fig. 5.17 shows the results of a gain and bandwidth measurement using a programmed cut-off frequency of 3 kHz (with no external high-pass filter and disabled DC-suppression). The input-referred offset voltage of the channels was additionally measured and varies between 34 μV and 12 μV within the programming range of the AFE.

The overall power dissipation of the system includes both, the analog circuits as well as the digital logic. In low-power mode and using a data rate of 500 Hz the

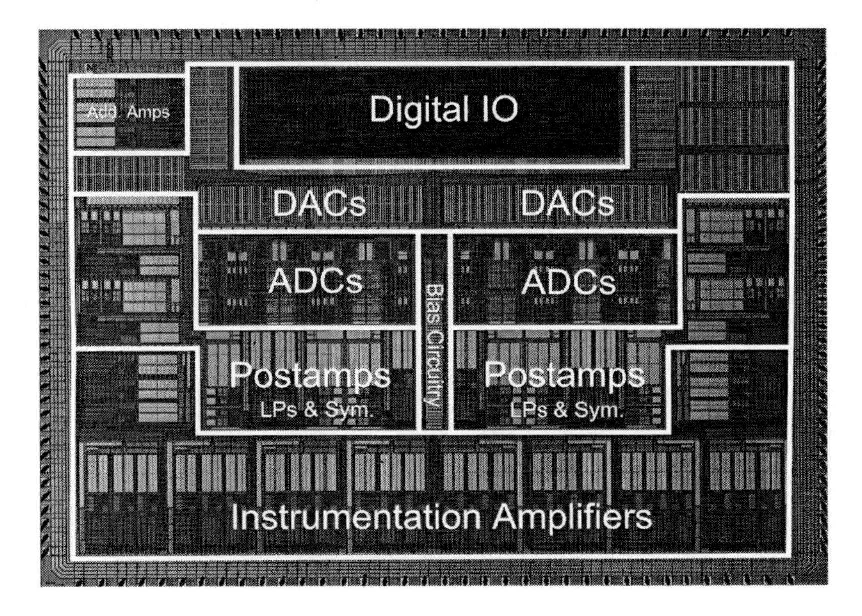

Figure 5.16: Microphotograph of the realized M3C3 system.

M3C3 has a power dissipation of 28.7 mW. Configuring the system to the high power mode and a data rate of 50 kHz results in a power consumption of 209.9 mW.

The M3C3 noise performance was measured using the digital output data. The data obtained therefore includes the noise contribution of the AFE and the 2nd-order $\Sigma\Delta$-modulator with the instrumentation amplifier of the AFE being the main noise source. The results of the noise measurements are listed in Table 5.7. The ECG noise constraints of Table 5.6 are always fulfilled, even for using the lowest power consumption and no chopper stabilization enabled. In case of EEG, the front-end has to be biased at a high power consumption with the chopper stabilization activated. The same argument holds for the EMG specification. In the case of EP, the measured noise exceeds the specified value slightly using a high power/chopper stabilization setup.

The noise performances calculated or simulated above match closely to the measured in the case of ECG as shown in Table 5.7. For the EEG noise including the complete AFE, the simulated value underestimates the measured results, however, the EEG design specifications are met in any case.

The NEF of just the M3C3 IA is estimated, like for the M3C2, by using the

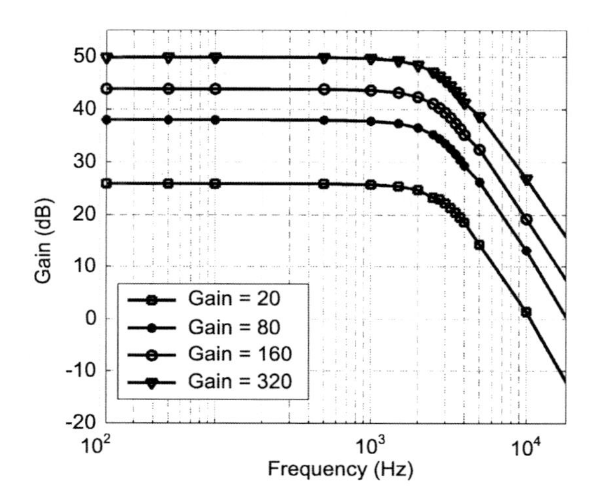

Figure 5.17: Measured gain and bandwidth of an analog channel, markers indicate measured data points.

Bandwidth (Signal type)	Condition	Experimental results	Specified	Calculated/ Simulated
0.05 Hz - 250 Hz (ECG)	$I_{bias,low}$ [1]	$1.13 - 1.74\,\mu$Vpp	$< 5\,\mu$Vpp	$1.48\,\mu$Vpp
	$I_{bias,low}$ [2]	$3.1 - 3.6\,\mu$Vpp		
0.05 Hz - 70 Hz (EEG)	$I_{bias,high}$ [1]	$< 0.68\,\mu$Vpp	$< 1\,\mu$Vpp	$0.44\,\mu$Vpp
0.01 Hz - 5 kHz (EMG)	$I_{bias,high}$ [1]	$< 3.9\,\mu$Vpp	$< 6\,\mu$Vpp	
0.1 Hz - 3 kHz (EP)	$I_{bias,high}$ [1]	$< 3.2\,\mu$Vpp	$< 3\,\mu$Vpp	

[1] Chopper stabilization enabled
[2] Chopper stabilization disabled

Table 5.7: Total input-referred voltage noise of the M3C3. The calculated values are from Section 5.3.2.

simulated IA power dissipation and taking the measured noise performance as an upper limit to the IA input-referred noise. For the ECG and EEG bandwidths, the NEF is given by 8.1 – 12.5 and 30, respectively. These values are somewhat higher than for the M3C2 IA (see Section 5.2.8) as a result of the additional op amp which is however needed to realize the DC-suppression circuit.

DC-suppression and CMRR Calibration

In order to test the DC-suppression circuit, the time constant of the on-chip integrator was set to 1 s and a 500 mVpp square-wave was applied to the V_{i+} input with V_{i-} connected to signal ground at $V_{sgnd} = 1.65$ V. Fig. 5.18 shows the applied square-wave and the output voltage of the IA, where it can be seen how the output voltage settles back to V_{sgnd} within the regulation time.

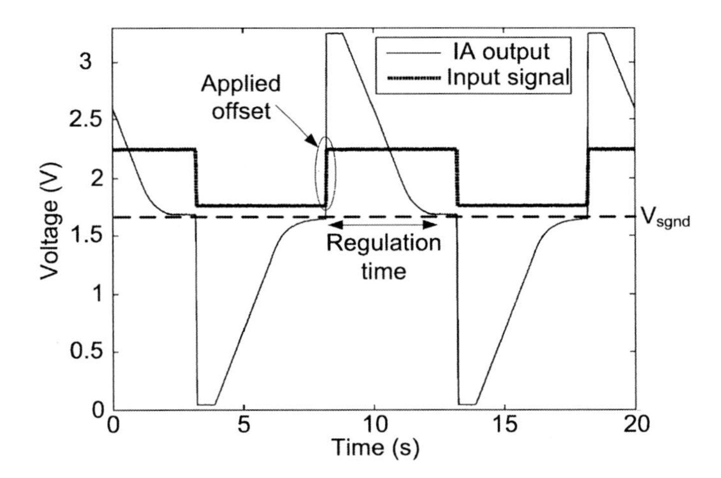

Figure 5.18: Example of the DC-suppression using a square-wave input. The IA output settles back to signal ground after one second.

The CMRR calibration was successfully utilized at medium and high power setting to obtain CMRR values of more than 80 dB at 50 Hz. A calibration value sweep for all 10 channels was performed at medium system power setting, Fig. 5.19a shows the corresponding CMRR values where at least one calibration step with a CMRR $>$ 80 dB can be found.

The CMRR was also tested for different common-mode frequencies with the AFE bias set to low and high power dissipation. The result at optimum calibration value

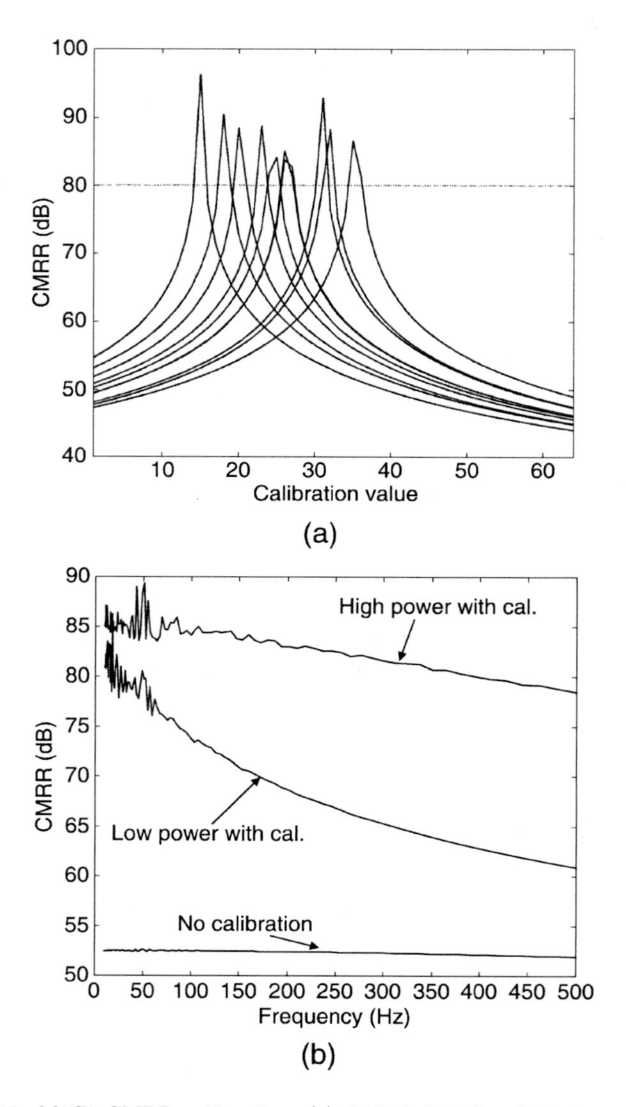

Figure 5.19: M3C3 CMRR calibration. (a) CMRR @ 50Hz of 10 channels versus calibration setup. (b) CMRR versus frequency in low and high power mode, and without calibration.

is shown in Fig. 5.19b. It can be seen that the CMRR at high power setting has a value of about 85 dB at 50 Hz, whereas for the low-power setup the CMRR just misses the 80 dB level. Both, the high and low-power setup CMRR curves drop with increasing frequency. In addition, a third curve is shown in Fig. 5.19b which gives the CMRR with no calibration applied.

The measurements shown in Fig. 5.19b confirm the general results of the 2-op amp CMRR analysis of Section 4.1.2 (neglecting the influence of the OP3 used for DC-suppression). The CMRR calibration reduces the effect of the resistor mismatch ($CMRR_{\Delta R}$) and the CMRR due to finite op amp open-loop gain ($CMRR_{\Delta OP}$) becomes the dominant effect. The difference between the CMRR values at low and high power setting is explained by the power-bandwidth trade-off of the programmable op amp architecture, i.e. at low I_{bias} the dominant pole of the op amp is moved to lower frequencies and the open-loop gain starts to drop from its DC level at lower frequencies. The influence of the open-loop gain on $CMRR_{\Delta OP}$ (see Section 4.1.2) can be generally seen by the decrease of CMRR for increasing frequencies. The curve showing the uncalibrated states reflects the frequency independence of the resistor mismatch related $CMRR_{\Delta R}$ value.

ADC and Digital Interface

The measurement of the ADC was performed by application of a single-tone test-signal and sweeping its amplitude. The resulting channel data was used to calculate the signal to noise and signal to noise and distortion ration (SNDR), Fig. 5.20 shows the measured SNR and SNDR versus input signal for a 70 Hz input tone and a data rate of 500 Hz. The results give an equivalent number of bits (ENOB) of 12.17 and 15.2 bits for the ECG and EEG bandwidth, respectively.

However, an anomaly in the linearity output data was found by application of a very low frequent triangle wave. The cause of the anomaly and its solution are explained in the next section.

The functionality of the digital interfaces was tested using an automated test environment developed to fully characterize the M3C3 system specifications [106]. A block diagram of the test system is depicted in Fig.5.21. It consist of a field programmable gate array (FPGA) implementing a serial control interface and serial high speed data interface. The system allows to control the setup of the M3C3 or read out its measured data on a personal computer using a USB interface. The maximum frequencies of operation are 100 kHz and 55 MHz for the serial control and data ports, respectively.

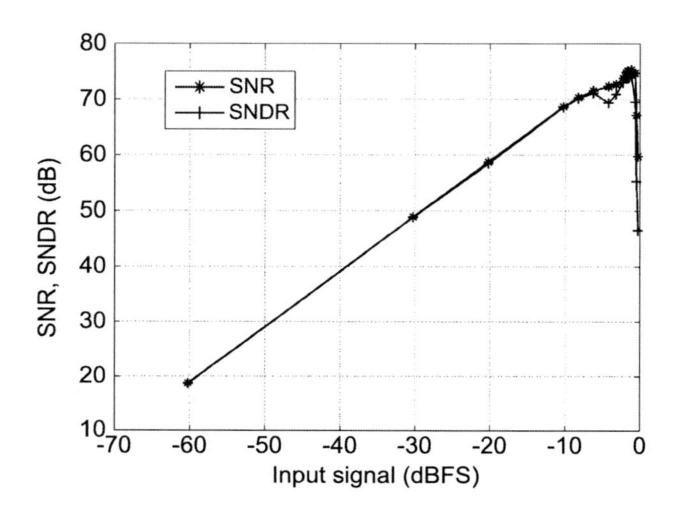

Figure 5.20: Measured SNR and SNDR versus input signal for a 70 Hz sine input and ECG bandwidth of 250 Hz.

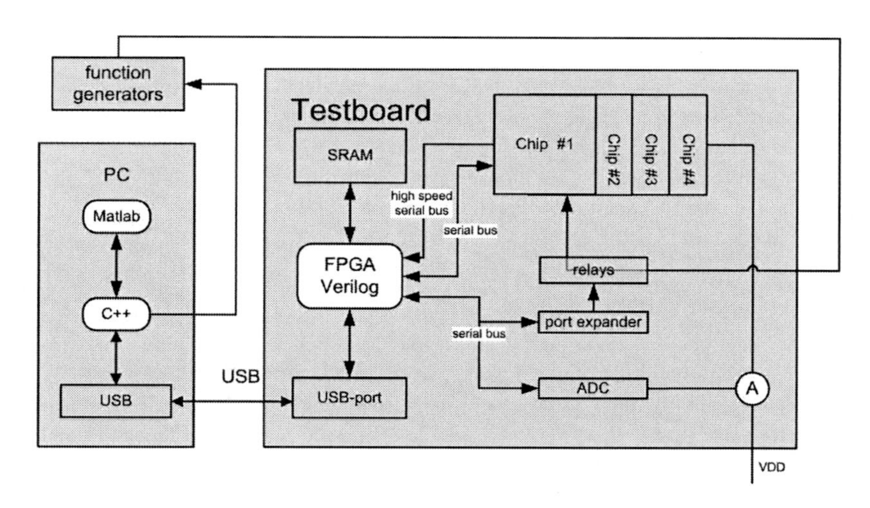

Figure 5.21: Test setup for digital communication.

Non-Linearity and Redesign

An anomaly in the linearity of the channel for an input range within the bottom half of the supply voltage range was discovered during the measurements. The non-linearity was investigated by means of simulating the stages following the IA up to the $\Sigma\Delta$-modulator input, as it could be ruled out that the problem is located within the IA or the ADC. A postlayout simulation of the PGA/low-pass/inverting amplifier blocks including resistors to model the global on-chip supply and ground paths was set up. In order to model the input resistance and switching effect of the $\Sigma\Delta$-modulator input circuit, an equivalent circuit using switches and appropriate sized resistors was used as a load circuit to the AFE output.

The simulation results showed that within the bottom half of the supply voltage range, the switching activity of the load circuit triggers small continuous oscillations within the signal path of the PGA/low-pass/inverting amplifier block. The frequency of the small oscillation slightly changes with increasing DC input voltage giving rise to the observed anomalies within the sampled channel data. An analysis of this behavior revealed that at the low-pass filter output a large parasitic capacitor of approximately 13 pF exists, see also Fig. 5.22.

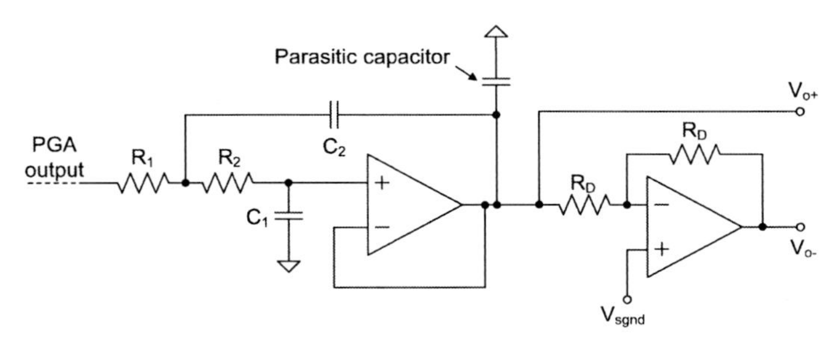

Figure 5.22: Location of the parasitic capacitor.

The parasitic capacitor is mainly resulting from the bottom plate of the Sallen-Key low-pass filter capacitor C_2. The additional capacitor increases the gain peaking of the low-pass filter op amp (which uses cascoded Miller compensation, see Section 4.2) to reach a value above 0 dB, thereby resulting in an instability of the amplifier. In order to solve this problem, the op amp was redesigned to eliminate any gain peaking by replacing its compensation circuit to the Miller compensation scheme also used in the OP130 designs of Section 4.4. Fig. 5.23 shows the above described postlayout simulation before (a) and after (b) the redesign.

The large spikes seen in this plot are a normal effect of the switching activity

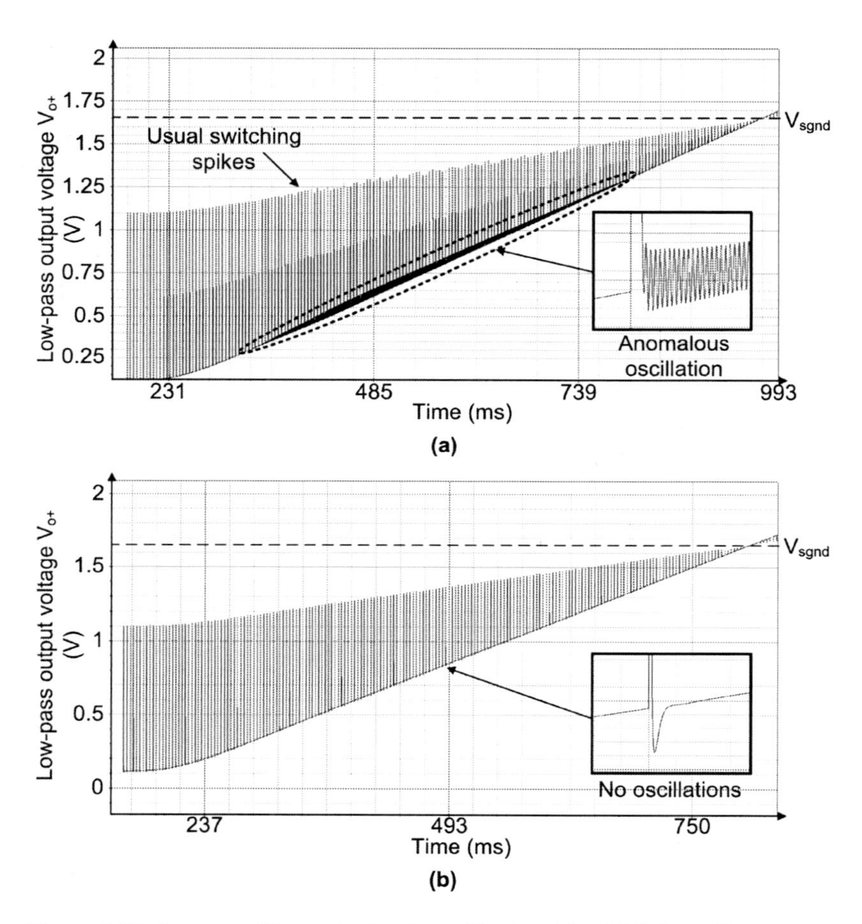

Figure 5.23: Low-pass filter output voltage V_{o+} (see Fig. 5.14) for a linear ramp input signal for the original design (a) and the redesign (b). Only the bottom half voltage range is shown. The spikes are usual and result from the switching activity of the load circuit. A zoomed view has been added for clarity.

by the ADC equivalent input circuit. It can been seen in Fig. 5.23b that the small oscillation disappeared completely after redesign. However, for the change of the compensation scheme the bias currents of the op amp had to be increased slightly which results in an additional 40 μA supply current for the analog front-end. A test chip including one original and one redesigned channel was sent for fabrication and is processed at the time of writing, hence no measurements can be given yet.

5.3.8 Results overview

The M3C3 presents a 10 channel biomedical signal acquisition system on a single chip with its main characteristics summarized in Table 5.8. The main target applications for the M3C3 are high density EEG recordings using more than 100 electrodes. For this, a multitude of M3C3s can be placed on one system board to realize the high number of available channels. Nevertheless, the M3C3 system can be also used for ECG, EMG and EP measurements with the possibility to realize mobile long-term ECG recording systems. To minimize board area, the system is encapsulated in a BGA package with a size of only 13 mm $\times$ 13 mm. A photo of the M3C3 in a BGA package with 144 pins and a Eurocent coin for comparison is shown in Fig. 5.24.

An example system for a 100 electrode EEG system could be realized on a double sided PCB with a dimension of 6 cm $\times$ 6 cm. The power consumption of the system would lie in a range of 1000 mW. A second possible application would be a mobile long-term ECG recording system with a size of a few cm^2. With the M3C3 in low-power configuration, two standard AA-batteries would be sufficient to power the system for more than one week.

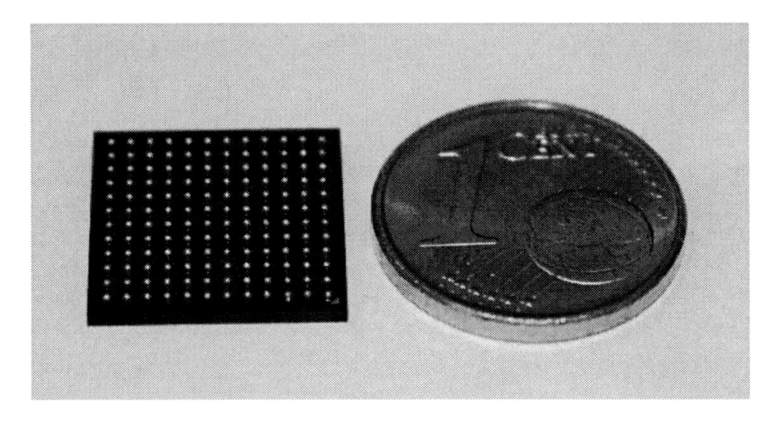

Figure 5.24: Photo of the M3C3 SoC in a BGA package.

Parameter	Condition/Remarks	Value
Supply voltage		3.3 V
Number of channels		10
Gain settings		20/80/160/320
Bandwidth	programmable	3/15 kHz
System power dissipation		28.7 - 209.9 mW
Total input-referred	ECG[2]	$\leq 1.74\mu$Vpp
noise[1]	EEG[3]	$\leq 0.68\mu$Vpp
	EMG[3]	$\leq 3.9\mu$Vpp
	EP[3]	$\leq 3.2\mu$Vpp
CMRR @ 50 Hz	using calibration $\geq$ 80 dB	
Input CM-range		$\pm$0.62 V
Offset comp. range	High-pass & DC-supp.	$\pm$1 V
Crosstalk	between channels	≤ -60 dB
Die area		47 mm^2
Pin number		144
Package	PGA/BGA	

[1] Vpp is calculated as the 6-fold value of the root-mean-square voltage noise Vrms
[2] Low-power mode
[3] Low-noise mode

Table 5.8: Main characteristics of the M3C3 biomedical SoC.

5.4 Conclusion

This chapter described the design and realization of two CMOS integrated biomedical signal acquisition systems using the front-end building block implementations of Chapter 4.

Each of the three channels of the first system (M3C2) exhibit an analog front-end that can be configured for low-noise or low-power operation and is followed by an analog to digital conversion. In addition, the system includes a digital signal processor, a respiration unit, and a built-in self test with on-chip analog front-end calibration. The M3C2 presents a system-on-chip (SoC) for biomedical signal acquisition with a wide range of functionality.

The second system (M3C3) incorporates ten channels each including a front-end and analog-to-digital conversion. The AFE includes CMRR calibration and a DC-suppression circuit. Like for the M3C2, the front-end is programmable with respect to noise and power performance. The input-referred DC-suppression range extends to a remarkable value of up to $\pm$1 V. The size of the realized M3C3 integrated circuit in a BGA package and its flexible, high-speed serial data output interface allows to build a high electrode count system using a multitude of M3C3s on a system board.

Chapter 6

Conclusion and Outlook

In this work, a systematic description of the design of CMOS amplifiers and front-end solutions for integrated biomedical signal acquisition systems is presented. The main focus of the design was to optimize the amplifiers (and resulting from this the analog front-ends) with respect to a low-noise and a low-power operation. The development of an operational amplifier in a 130 nm process technology for use in a biomedical signal acquisition front-end was of particular importance as it shows the feasibility using deep-submicron process technologies in high precision analog circuits.

In order to obtain a comprehensive description of the biomedical amplifiers and front-end designs, it is necessary to treat the fundamentals as well as the applications. The treatment of the basics involves two different fields that have to be covered: The fundamentals of biological and medical systems and the basics of semiconductors and circuit design. Both fundamentals have been outlined in the first two chapters.

Furthermore, application examples are essential to demonstrate the utilization of the amplifiers and front-end designs with respect to electrical and biomedical constraints in the context of complete system solutions. Therefore, the design and realization of two integrated system solutions including several channels for the acquisition of biomedical signals as well as digital interface circuitry was described in the last chapter of this work. Each of the system channels exhibits a programmable front-end and an ADC.

6.1 Main Findings and Discussion

The main finding of this work are summarized in the following list:

1. Chapter 4: A systematic description of the main front-end building blocks characteristics including instrumentation amplifiers, programmable gain amplifiers and low-pass filters was presented. The characteristics discussed include noise performance, power dissipation and additional building block specific parameters. An outstanding circuit specific parameter is the common-mode rejection ratio of the 2-op amp instrumentation amplifier.

   The detailed analysis of this parameter was motivated by the use of the 2-op amp IA topology for the M3C2 and M3C3 integrated biomedical signal acquisition systems with no detailed analysis available in literature. It was shown that the CMRR of the 2-op amp IA is made of three components: The first related to resistor mismatch, the second related to the open loop gain of the bottom op amp and the third related to the difference in CMRR between both op amps.

   The resistor related CMRR is usually the dominant effect and masks the other two. However, these have to be considered if a good resistor matching is achieved by means of using a calibration technique. In this case, the CMRR of the 2-op amp IA equals approximately the open loop gain of the bottom op amp. Moreover, the frequency dependence of the op amp open loop is reflected by a frequency dependence of the 2-op amp IA CMRR.

2. Chapter 4: The use of CMOS op amps that are programmable with respect to a noise-power trade-off allows the design of analog front-ends that can be set to an optimal specification regarding the biomedical signals applied. Both an existing 350 nm and a newly developed 130 nm programmable op amp design were presented. The 130 nm programmable op amp design presents several innovative features compared to op amp designs in conservative CMOS process technologies.

   First, it was shown that the use of multi-threshold processes allows to keep circuit topologies initially developed for CMOS processes with higher supply voltages in deep-submicron process technologies. Second, the techniques to enable the programmability of the op amp were successfully applied for a 130 nm design. In this context, the existing systematic design methodology for a cascoded Miller compensated programmable op amp was extended to the Miller compensated case. Third, a novel constant-gm circuit was developed that allows to use a current switch solution within the op amp input circuit which is modified to enable the programmability.

In order to compare the 130 nm op amp design to other work in this field Table 6.1 has been compiled. Unfortunately, no noise performance has been given for the op amp designs in a 120/130 nm technology. Therefore, this table comprises also designs using a 180 nm process technology. It can be seen that the noise performance of the proposed 130 nm op amp design is outperformed only by the design in [107] which exhibits, however, a very low DC open loop gain of 53 dB. This value will most probably decrease within the 130 nm technology node considering that the intrinsic gain reduces for smaller technologies. When programmed to be in low-power mode, the presented design exhibits the lowest power dissipation within the 120 nm designs of [86] and [108].

A comparison of all designs in Table 6.1 regarding the power consumption shows that [109] exhibits the lowest power consumption. However, this extremely low-power consumption is associated with a very high thermal voltage noise density of 120 $\mathrm{nV}/\sqrt{\mathrm{Hz}}$.

3. Chapter 5: The presented M3C2 and M3C3 system solutions demonstrate successfully the implementation of the analog front-end design building blocks described in Chapter 4.

   The three channel M3C2 chip presents an innovation with respect to the design of biomedical signal acquisition systems by integrating a multitude of different analog and digital components into one system-on-chip. The integrated sub-systems include a programmable analog front-end, a build-in self test, an oscillator, a CMRR calibration, an on-chip digital signal processor and a respiration measurement unit. The latter introduces a novelty in integrated biomedical solutions by fully integrating a respiration monitoring system within an ECG recording system. The applicability of this system was shown by an experimental mobile EEG system and a portable battery powered ECG system.

   The second system proposed, M3C3, includes ten channels, each equipped with a programmable analog front-end similar to that of the M3C2. However, this system solution introduces a new CMRR calibration scheme by exploiting a switchable resistor array. Additionally, a modified 2-op amp IA topology introduces a novel DC-offset suppression scheme with an input DC-offset range of up to $V_{DCsuppr} = \pm 1000$ mV.

   In order to compare the performance of the DC-offset suppression circuit, a

Reference	**This work**	[86]	[108]	[110]	[109]	[107]
Tech. (nm)	130	120	120	180	180	180
V_{DD} (V)	1.2	1.5	$0.8 - 1.4$	1.5	0.8	1.2
P_{OP} (mW)	$0.2 - 2.5$	1.8	$0.4 - 0.9$	$0.1 - 0.16$	0.006	0.2
$v_{nith,OP}$ (nV/$\sqrt{\text{Hz}}$)	9[1]– 5.1	-	-	$15.9 - 24.5$	120	0.8
A_{d0} (dB)	$79 - 76.5$	73	$63.5 - 80$	70.6/73.6	74	52
f_u (MHz)	$1.3 - 8.9$	4.36	$12.3 - 20.5$	6 / 7.5	0.87	1.9
PM (o)	$62 - 58$	70	$49 - 54.4$	64	66	92
A_{OP} (mm^2)	0.068[2]	0.01	0.024	-	0.033	0.03
CMRR (dB)	$64 - 72$[2]	57.3	74	-	75	-
R-to-R (I/O)[4]	y/y	y/y	n/y	y/y	y/y	y/y
Load C_L/R_L (pF/kΩ)	13.5/22	5/-	10/750	5/1000	12/100	28/50
Note	Prog.	Fully diff.		Sim. only		

[1] simulated value
[2] OP130b
[3] for $V_{i,vm} = V_{SS} + 0.2\text{V} \ldots V_{DD} - 0.2\text{V}$
[4] R-to-R (I/O): Rail-to-Rail (Input/Output), y: yes, n: no

Table 6.1: Performance comparison of deep-submicron op amp designs.

figure of merit can be introduced by relating the input DC-offset range using DC-suppression to the standard input DC-offset range of an amplifier. The maximum input DC-offset range of a standard, DC-coupled instrumentation amplifier is given by the supply voltage V_{DD} divided by the gain of the amplifier A. An *input DC-suppression to standard DC-offset range ratio (DSOR)* can be now defined as

$$\text{DSOR} = A\frac{V_{DCsuppr}}{V_{DD}} \tag{6.1}$$

where an DSOR of one is obtained if no DC-suppression is used.

The DC-offset performance of different systems including data from a comparison given in [91], the M3C3 and two additional state-of-the-art biomedical signal acquisition systems is shown in Fig. 6.1. It can be seen that M3C3 achieves the second highest DSOR value and by far the highest absolute input DC offset range.

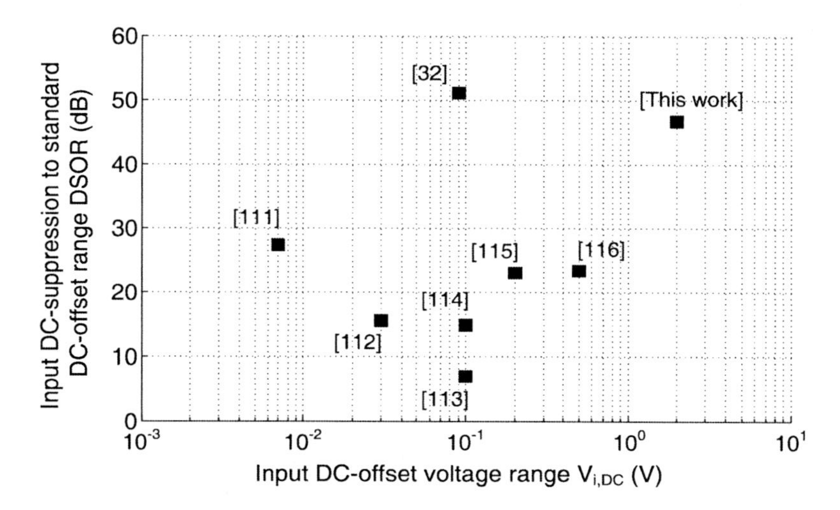

Figure 6.1: Comparison of DC offset suppression capabilities: DSOR versus input DC-offset range for cited publications and the M3C3.

6.2 Future Work

The future work recommended includes a redesign of the OP130 constant-gm circuit to further reduce the transconduction variation. A first step towards this would be the measurement of the input and cascode circuit only with respect to the performance of the constant-gm circuit. This would exclude the limitations given by the output transistors going into the linear region when the op amp output voltage reaches the supply rails. However, this requires a chip realization of the input and cascode circuit only.

In a second step, a circuit solution has to be found to minimize the V_{DS} variation of the input pairs. This would limit the large change in input pair output resistance. Further research is needed with respect to novel circuit techniques to allow the design of high performance operational amplifiers for technology nodes down to a few nanometers.

The trend towards implanted medical devices necessitates the design of advanced integrated signal acquisition circuit systems and in particular the development of novel circuit topologies to meet the constraints of implanted systems. For such systems, additional design aspects become important, namely the system size, the packaging and the biocompatibility of the system. The size of implantable systems can be reduced significantly using novel chip packaging solutions like system-in-package (SiP) which incorporate several dies within one package or by attaching and bonding the die directly on the system board. The biocompatibility of the system to be implanted requires a research towards novel packaging materials. A low-power dissipation of such systems is essential in order to minimize the heating of the tissue surrounding the implant and to allow inductive powering as well as wireless transmission of data.

Appendix A

IA Calculations

In the following IA calculations, two substitutions related to the IA input voltages will be used to simplify the derivation of the equations:

- The positive and negative IA input voltage V_{i+} and V_{i-} will be related to a single input voltage V_i given by

$$V_{i+} - V_{i-} = \frac{+V_i}{2} - \frac{-V_i}{2} = V_i$$

- The gain of the IA is divided into the differential and common-mode gain. The input differential voltage is given by $(V_{i+} - V_{i-})$ and the input common-mode voltage by $(V_{i+} + V_{i-})/2$, hence

$$V_o = A_d\,(V_{i+} - V_{i-}) + A_{cm}\,\frac{V_{i+} + V_{i-}}{2}$$

  with A_d and A_{cm} being the differential and common-mode gain, respectively, and V_o is the output voltage.

A.1 3-Op Amp IA

A.1.1 Gain

The ideal gain of the 3-op amp IA which is depicted in Fig. A.1 is given by

$$A_{3IA} = 1 + 2\,\frac{R_1}{R_2} \tag{A.1}$$

if $R_1 = R_3$ and $R_4 = R_5 = R_6 = R_7$.

To determine the gain error due to resistor mismatch we first consider the gain of the difference amplifier consisting of OP_3 and associated resistors $R_4 \ldots R_7$. A

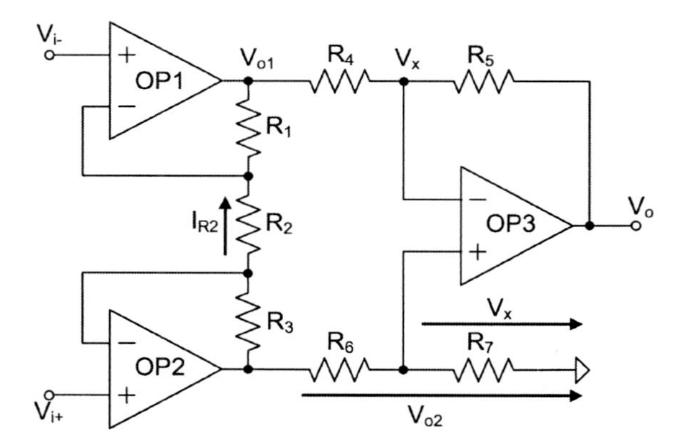

Figure A.1: 3-op amp IA topology

circuit analysis of the difference amplifier gives

$$\frac{V_x}{V_{o2}} = \frac{R_7}{R_6 + R_7} \tag{A.2}$$

$$\frac{V_{o1} - V_x}{R_4} = \frac{V_x - V_o}{R_5} \tag{A.3}$$

Solving (A.2) for V_x and substituting this result in (A.3) gives

$$V_o = \frac{R_7 (R_4 + R_5)}{R_4 (R_6 + R_7)} V_{o2} - \frac{R_5}{R_4} V_{o1} \tag{A.4}$$

$$= \frac{R_4 + R_5}{R_4 (R_6/R_7) + R_4} V_{o2} - \frac{R_5}{R_4} V_{o1} \tag{A.5}$$

$$\tag{A.6}$$

The relative mismatch between the gain setting resistors is now defined as

$$\frac{R_7}{R_6} = (1 \pm \Delta_{3IA_3}) \frac{R_5}{R_4} \tag{A.7}$$

with Δ_{3IA_3} as the relative mismatch between both resistor ratios. Using (A.7) and assuming a unity gain difference amplifier, i.e. $R_5 = R_4$ (A.5) becomes

$$V_o = \frac{2}{\left(\frac{1}{1 \pm \Delta_{3IA_3}}\right) + 1} V_{o2} - V_{o1}. \tag{A.8}$$

The gain of difference amplifier A_3 is now determined by defining $V_{o1} = -V_i/2$ and $V_{o2} = V_i/2$ and is hence given by

$$A_3 = \frac{V_o}{V_i} = 1 - \frac{1}{2} + \frac{1 \pm \Delta_{3IA_3}}{2 \pm \Delta_{3IA_3}} \tag{A.9}$$

We now consider the gain error associated with the first stage of the 3-op amp IA. For this we first calculate the current in R_2 which is given by

$$I_{R2} = \frac{V_{i+} - V_{i-}}{R_2}. \tag{A.10}$$

This current flows also through resistors R_1 and R_3, hence

$$\frac{V_{i+} - V_{i-}}{R_2} = \frac{V_{o2} - V_{o1}}{R_1 + R_2 + R_3} \tag{A.11}$$

$$\Rightarrow A_{12} = \frac{V_o}{V_{i+} - V_{i-}} = \frac{R_1 + R_2 + R_3}{R_2} \tag{A.12}$$

The relative mismatch between the gain setting resistor ratios is now given by

$$\frac{R_3}{R_2} = (1 \pm \Delta_{3IA_{12}})\frac{R_1}{R_2} \tag{A.13}$$

and the first stage gain becomes accordingly

$$A_{12} = 1 + 2\frac{R_1}{R_2} \pm \Delta_{3IA_{12}}\left(\frac{R_1}{R_2}\right) \tag{A.14}$$

Multiplying (A.9) and (A.14) results in the overall 3-op amp IA gain including resistor mismatch, it is given by

$$A_{\Delta 3IA} = A_{12} \cdot A_3 = \Bigg(\underbrace{1 + 2\frac{R_1}{R_2}}_{A_{3IA}} \underbrace{\pm \Delta_{3IA_{12}}\left(\frac{R_1}{R_2}\right)}_{\Delta A_{12}}\Bigg)\Bigg(1 \underbrace{- \frac{1}{2} + \frac{1 \pm \Delta_{3IA_3}}{2 \pm \Delta_{3IA_3}}}_{\Delta A_3}\Bigg) \tag{A.15}$$

A.2 2-Op Amp IA

A.2.1 Gain

The 2-op amp IA gain is given as

$$A_{2IA} = \frac{V_o}{V_i} = 1 + \frac{R_1}{R_2} \tag{A.16}$$

for ideal resistor matching, i.e. $R_4 = R_1$ and $R_3 = R_2$ in Fig. A.2. To determine the influence of resistor ratio mismatch on the gain error we set

$$\frac{R_4}{R_3} = (1 \pm \Delta_{2IA})\frac{R_1}{R_2} \tag{A.17}$$

with Δ_{2IA} being the relative mismatch between both resistor ratios. The 2-op amp gain V_o/V_i can be calculated analyzing Fig. A.2. From this we obtain two equations, with the first one being the output voltage of the lower non-inverting amplifier, i.e.

$$V_x = \frac{-V_i}{2}\left(1 + \frac{R_2}{R_1}\right) \tag{A.18}$$

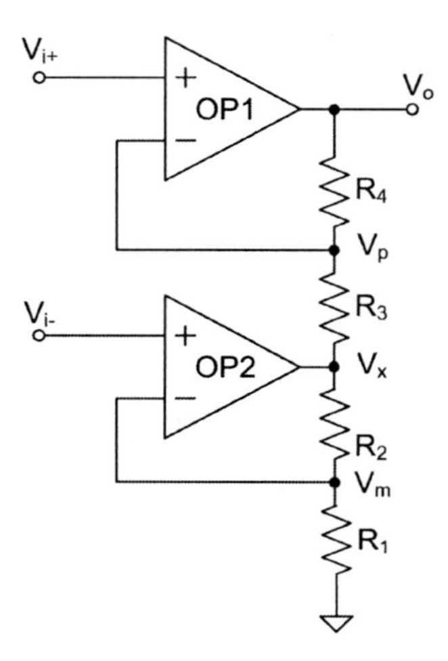

Figure A.2: 2-op amp IA

and the second one equating the currents in resistors R_3 and R_4 which results in

$$\frac{V_o - \frac{V_i}{2}}{R_4} = \frac{\frac{V_i}{2} - V_x}{R_3} \tag{A.19}$$

Using (A.18) and (A.19) we find

$$A_{2IA} = \frac{V_o}{V_i} = \frac{1}{2}\left[1 + 2\frac{R_4}{R_3} + \frac{R_2 R_4}{R_1 R_3}\right] \tag{A.20}$$

If we now include the resistor mismatch relation (A.17) we finally arrive at the overall gain A_{2IA} which includes resistor mismatch:

$$A_{\Delta 2IA} = \underbrace{\left(1 + \frac{R_1}{R_2}\right)}_{A_{2IA}} \underbrace{\pm \Delta_{2IA}\left(\frac{1}{2} + \frac{R_1}{R_2}\right)}_{\Delta A_{2IA}}. \tag{A.21}$$

A.2.2 CMRR

For all CMRR calculation we use the circuit of Fig. A.3 to obtain the common-mode voltage gain A_{cm} by setting $v_p = v_m = v_i$ and calculating v_o/v_i. First, we just consider resistor mismatch and assume the op amps to be ideal. The common-mode gain $A_{cm,\Delta R}$ is hence given by

$$\frac{v_o}{v_i} = 1 - \frac{R_4 R_2}{R_1 R_3} = A_{cm,\Delta R} \tag{A.22}$$

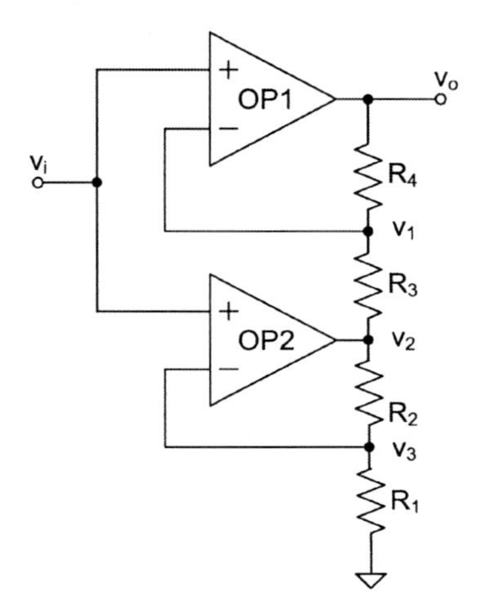

Figure A.3: 2-op amp IA CMRR Measurement

or

$$A_{cm,\Delta R} = 1 - \frac{R_1 R_2}{R_2 R_1}(1 \pm \Delta_{2IA}) = \mp\Delta_{2IA} \tag{A.23}$$

by including the resistor ratio mismatch relationship (A.17). The common-mode rejection ratio due to resistor mismatch becomes accordingly

$$\mathrm{CMRR}_{\Delta R} = \frac{A_{2IA}}{\mp\Delta_{2IA}}. \tag{A.24}$$

In order to include non-ideal op amps we assume the op amps to have an open-loop gain of A_d and a common-mode gain of A_{cm}. Examining Fig. A.3 with respect to voltages v_o and v_2 and to the currents in resistors R_3, R_4 and R_2, R_1 we find the following system of equations:

$$v_o = A_{d1}(v_i - v_1) + \frac{A_{cm1}}{2}(v_i + v_1) \tag{A.25}$$

$$v_2 = A_{d2}(v_i - v_3) + \frac{A_{cm2}}{2}(v_i + v_3) \tag{A.26}$$

$$\frac{v_o - v_1}{R_1} = \frac{v_1 - v_2}{R_2} \tag{A.27}$$

$$\frac{v_2 - v_3}{R_2} = \frac{v_3}{R_1} \tag{A.28}$$

(A.29)

Solving this system of equations for v_o/v_i gives

$$\frac{v_o}{v_i} = \frac{2(R_1+R_2)[2A_{d1}(R_1+R_2)+A_{cm1}(R_1+R_2)+2R_1(A_{d2}A_{cm1}-A_{d1}A_{cm2})]}{[2(R_1+R_2+A_{d2}R_1)-A_{cm2}R_1][2(R_1+R_2+A_{d1}R_2)-A_{cm1}R_2]} \tag{A.30}$$

which can be simplified to

$$\frac{v_o}{v_i} = \frac{(R_1+R_2)[A_{d1}(R_1+R_2)+R_1A_{d1}A_{d2}(1/\text{CMRR}_{OP1}-1/\text{CMRR}_{OP2})]}{(R_1+R_2+A_{d2}R_1)(R_1+R_2+A_{d1}R_2)} \tag{A.31}$$

by assuming $A_{cm1,2} \cdot R_{1,2} \approx 0$ and setting $A_{cm} = A_d/\text{CMRR}$. If we further assume that $A_{d1,2} \cdot R_{1,2} \gg R_{1,2}$ we arrive at the more clearly form of

$$\frac{v_o}{v_i} = \underbrace{\frac{(R_1+R_2)^2}{A_{d2}R_1R_2}}_{A_{cm,\Delta OP}} + \underbrace{\left(1+\frac{R_1}{R_2}\right)\left(\frac{1}{\text{CMRR}_{OP1}} - \frac{1}{\text{CMRR}_{OP2}}\right)}_{A_{cm,\Delta CMRR}} \tag{A.32}$$

where $A_{cm,\Delta OP}$ denotes the common-mode gain influenced by the bottom op amp open-loop gain and $A_{cm,\Delta CMRR}$ a common-mode gain resulting from CMRR difference between both op amps.

The overall common-mode rejections ratio is obtained by superposition of the reciprocals of the separate common-mode rejection ratios [39]. Accordingly, using (A.22), (A.23) and (A.32) one obtains

$$\frac{1}{\text{CMRR}_{2IA}} = \frac{A_{cm,\Delta OP}}{A_{2IA}} + \frac{A_{cm,\Delta CMRR}}{A_{2IA}} + \frac{A_{cm,\Delta R}}{A_{2IA}} \tag{A.33}$$

$$= \frac{1}{\text{CMRR}_{\Delta OP}} + \frac{1}{\text{CMRR}_{\Delta CMRR}} + \frac{1}{\text{CMRR}_{\Delta R}} \tag{A.34}$$

The 2-op amp IA is now considered with respect to CMRR_{OP2} only and the op amps are modeled as a simple one pole low-pass with transfer function

$$A_d = A_{d0}/(1+j(\omega/\omega_p)) \tag{A.35}$$

where ω_p denotes the frequency of the pole and A_{d0} is the DC open-loop gain. Now, $\text{CMRR}_{\Delta OP}$ becomes also frequency dependent and $\text{CMRR}_{\Delta OP}$ can be written as

$$\text{CMRR}_{\Delta OP} = \frac{A_{2IA}}{A_{cm,OP2}} = \frac{1}{A_2} \cdot \left|\frac{A_{d0}}{1+j(\omega/\omega_{p1})}\right| \tag{A.36}$$

where $A_2 = 1 + R_2/R_1$ denotes the gain of the lower op amp. Equation (A.36) describes the 2-op amp IA AC CMRR characteristic. Considering $\text{CMRR}_{\Delta CMRR}$ only results in

$$\text{CMRR}_{\Delta CMRR} = \frac{A_{2IA}}{A_{cm,\Delta CMRR}} = \left(\frac{1}{CMRR_1} - \frac{1}{CMRR_2}\right)^{-1}. \tag{A.37}$$

A.2.3 Noise

The 2-op amp IA noise is calculated by adding voltage noise sources to the resistors and to the op amps as depicted in Fig. 4.7. The resistors exhibit a voltage noise PSD of $v^2_{n,R}(f) = 4kTR$ and $v^2_{n,OP}(f)$ includes all op amp noise sources (thermal and 1/f) referred to the positive input terminal.

The contribution of each noise source to the output noise PSD $v^2_{no,2IA}(f)$ is given by

$$v^2_{no,OP1}(f) = v^2_{n,OP1}(f)\left(1+\frac{R_4}{R_3}\right)^2 = v^2_{n,OP1}(f)\left(1+\frac{R_1}{R_2}\right)^2 \quad \text{(A.38)}$$

$$v^2_{no,OP2}(f) = v^2_{n,OP2}(f)\left(1+\frac{R_2}{R_1}\right)^2\left(\frac{R_4}{R_3}\right)^2 = v^2_{n,OP2}(f)\left(1+\frac{R_1}{R_2}\right)^2 \quad \text{(A.39)}$$

for the op amps and by

$$v^2_{no,R1}(f) = v^2_{n,R1}(f)\left(\frac{R_2}{R_1}\right)^2\left(\frac{R_4}{R_3}\right)^2 = v^2_{n,R1}(f) \quad \text{(A.40)}$$

$$v^2_{no,R2}(f) = v^2_{n,R2}(f)\left(\frac{R_4}{R_3}\right)^2 = v^2_{n,R2}(f)\left(\frac{R_1}{R_2}\right)^2 \quad \text{(A.41)}$$

$$v^2_{no,R3}(f) = v^2_{n,R3}(f)\left(\frac{R_1}{R_2}\right)^2 \quad \text{(A.42)}$$

$$v^2_{no,R4}(f) = v^2_{n,R4}(f) \quad \text{(A.43)}$$

for the resistors. Assuming that the op amps have the same input-referred noise and that $R_4 = R_1$ and $R_3 = R_2$ we can write

$$v^2_{n,OP}(f) = v^2_{n,OP1}(f) = v^2_{n,OP2}(f) \quad \text{(A.44)}$$

$$v^2_{n,R4}(f) = v^2_{n,R1}(f) \quad \text{(A.45)}$$

$$v^2_{n,R2}(f) = v^2_{n,R3}(f). \quad \text{(A.46)}$$

The sum of the noise PSD of each source to the output results in the total output noise PSD $v^2_{no,2IA}(f)$ which is given by

$$v^2_{no,2IA}(f) = 2v^2_{n,R1}(f) + 2\left(\frac{R_1}{R_2}\right)^2 v^2_{n,R2}(f) + 2\left(1+\frac{R_1}{R_2}\right)^2 v^2_{n,OP}(f) \quad \text{(A.47)}$$

Using the the relation

$$\frac{v^2_{n,R2}(f)}{v^2_{n,R1}(f)} = \frac{4kTR_2}{4kTR_1} \Leftrightarrow v^2_{n,R2}(f) = v^2_{n,R1}(f)\left(\frac{R_2}{R_1}\right) \quad \text{(A.48)}$$

a more simple form of (A.47) is obtained by

$$v_{no,2IA}^2(f) = 2\left[\underbrace{\left(1+\frac{R_1}{R_2}\right)}_{A_{2IA}} v_{n,R1}^2(f) + \underbrace{\left(1+\frac{R_1}{R_2}\right)^2}_{A_{2IA}^2} v_{n,OP}^2(f)\right]. \tag{A.49}$$

Dividing this result by the 2-op amp IA gain gives the input-referred noise PSD:

$$v_{ni,2IA}^2(f) = 2\left(\frac{v_{n,R1}^2(f)}{A_{2IA}} + v_{n,OP}^2(f)\right). \tag{A.50}$$

Appendix B

Op Amp Measurement Setups

This Appendix describes the measurement setups used for the OP130a and OP130b designs of Section 4.4. The setups are mostly based on the op amp measurement circuits given in [76] and [117]. Some of the measurement setups used differ for the OP130a and OP130b, hence, two test procedures are presented in such a case. The op amp being tested will be labeled *AUT (amplifier under test)* in the following schematics. Fig. B.1 shows the basic setup regarding supply voltage, signal ground and bias currents.

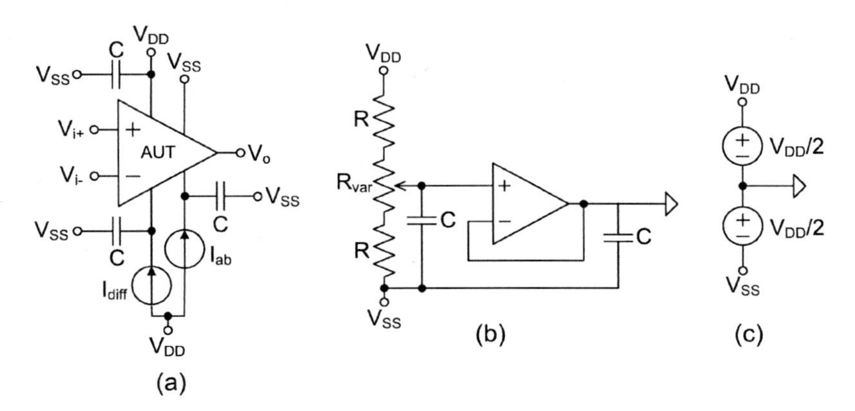

Figure B.1: Bias currents and supply/signal ground voltage generation. (a) Op Amp circuit including bias current generation. (b) signal ground generation for the OP130a. (c) Supply voltages and signal ground of the OP130b.

The generation of the variable bias currents I_{diff} and I_{ab} in Fig. B.1a was realized for the OP130a using two resistor arrays, where each array includes seven resistors that are all connected at one terminal to V_{DD}. The other terminal of each resistor can be connected separately via jumpers to the I_{diff} and I_{ab} input terminals of the op amp, respectively. The resistor values were chosen such, that the currents generated have a value as given in Table 4.3. In the case of the OP130b, a Keithley 220 programmable current source and a Keithley 6430 sub-femtoamp sourcemeter [118] were used to generate the bias currents directly. For the OP130a measurements, signal ground was generated using a resistive voltage divider using $R = 100$ kΩ, $R_{var} = 1$ kΩ and an op amp configured as a voltage follower as shown in Fig. B.1b. The OP130b setups use two supply voltages, each 0.6 V, referenced with respect to signal ground as depicted in Fig. B.1b. The capacitors in Fig. B.1 suppress the noise and buffer the supply and signal ground voltages (for the OP130a capacitors with $C = 10\,\mu$ F and $C = 10\,\mu\text{F}||10nF$, respectively, were used).

The supply voltage of 1.2 V was generated by voltage regulators for the OP130a measurements. A Keithley 4200 SCS Semiconductor Characterization System was used to generate 2 × 0.6 V for the OP130b test. An exception to this were battery powered noise measurements in order to minimize any additional noise sources.

B.1 Static Power Dissipation

The static power dissipation of the OP130a and OP130b was determined by measuring the supply current and adding the power consumption due to the bias currents I_{diff} and I_{ab}. The OP130a static supply current measurement is shown in Fig. B.2a.

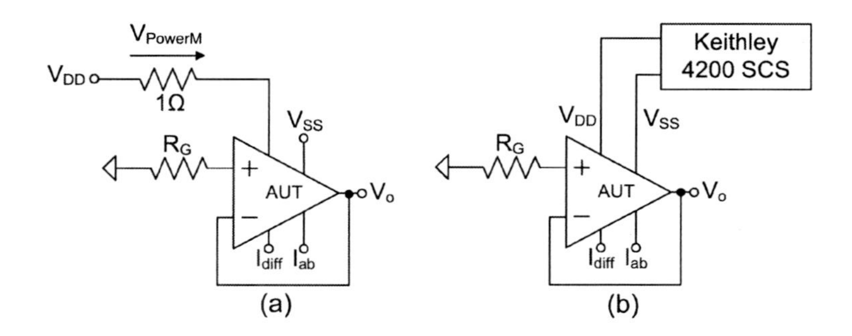

Figure B.2: Power dissipation measurement. (a) Measurement of the supply current using a 1 Ω resistor. (b) Supply voltage and supply current measurement using a parameter analyzer.

A 1 Ω resistor was added in the positive supply voltage path and the voltage

V_{PowerM} across this resistor was measured. The AUT was configured as a voltage follower with the input terminal of the AUT connected to signal ground via R_G ($= 50\,\Omega$). The total static power dissipation is given by

$$P_{OP,stat} = 1.2\ \mathrm{V}\left(\frac{V_{PowerM}}{1\,\Omega} + I_{diff} + I_{ab}\right) \tag{B.1}$$

For the OP130b, the Keithley 4200 SCS parameter analyzer was utilized to generate the supply voltages and to measure the supply current simultaneously. The total static power consumption is determined by replacing the term $V_{PowerM}/1\,\Omega$ in (B.1) by the measured current.

B.2 Noise Performance

A Rohde & Schwarz FSIQ 26 Signal Analyzer [119] was used to measure the AUT noise, Fig. B.3 shows the complete setup. The AUT is configured as a non-inverting amplifier. For the OP130a noise measurement, a gain of 101 was realized using $R_1 = 10\ \mathrm{k}\Omega$ and $R_2 = 100\,\Omega$ and for additional amplification of the noise, an additional amplifier having a gain of 10 was used. The OP130b measurements were performed with no additional amplification.

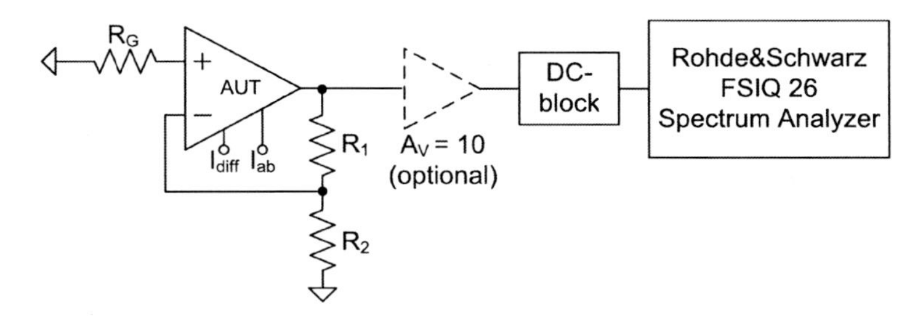

Figure B.3: Noise measurement using a spectrum analyzer.

In order to protect the input circuitry of the spectrum analyzer from any DC voltage applied (this can damage the input mixers), an additional DC-block was used as a high-pass filter with a cut-off frequency in the range of the start frequency of the spectrum analyzer (20 Hz). The measured voltage noise density was divided by the gain of 101 (or 1010 when using the additional amplifier) in order to obtain the input-referred noise PSD $v_{ni,OP}^2(f)$.

B.3 DC Open Loop Gain

The DC open-loop gain measurement setup is based on a feedback loop using an auxiliary amplifier as shown in Fig. B.4.

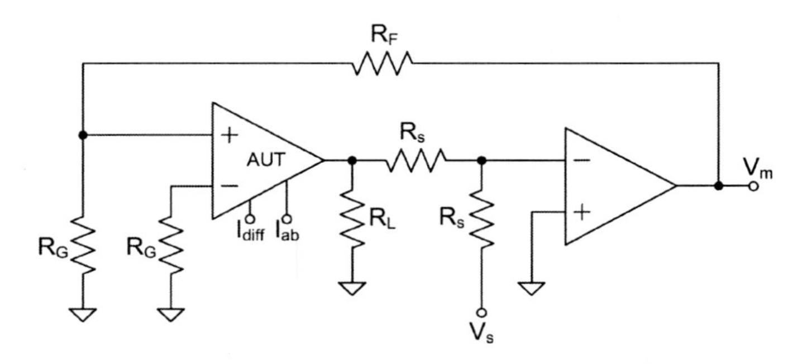

Figure B.4: Open-loop gain measurement setup.

The output voltage of the auxiliary amplifier, V_m, is attenuated by the voltage divider formed by R_F/R_G and forces the AUT to have an output voltage corresponding to the applied voltage V_s. The voltage V_s was set to $V_{s1} = -500$ mV, $V_{s2} = 0$ V and $V_{s3} = +500$ mV which resulted in measured voltages V_{m1}, V_{m2} and V_{m3}. The DC open-loop gain with respect to a positive or negative excursion, A_{d0+} and A_{d0-}, was calculated using

$$A_{d0+} = \frac{|V_{s1} - V_{s2}|}{|V_{m1} - V_{m2}|}\frac{R_F}{R_G} \tag{B.2}$$

$$A_{d0-} = \frac{|V_{s2} - V_{s3}|}{|V_{m2} - V_{m3}|}\frac{R_F}{R_G} \tag{B.3}$$

For the OP130b measurements, resistor values of $R_F = 100\,\mathrm{k\Omega}$, $R_G = 50\,\Omega$ and $R_s = 100\,\mathrm{k\Omega}$ were used. An average DC open-loop gain value, A_{d0}, is obtained by taking the mean value of both, the positive and negative open-loop gain results. The OP130b DC open-loop measurements are essentially consistent with the OP130a results using the simple approach described in Section 4.4.5.

B.4 Unity Gain Frequency and Phase Margin

In order to measure the unity gain frequency and the phase margin, the AUT was configured in a non-inverting amplifier configuration with the feedback resistors R_1

and R_2 setting the gain of the amplifier to 101. The unity gain frequency and the phase margin of the AUT were measured using a Rohde & Schwarz ZVRE network analyzer. The network analyzer exhibits a 50 Ω input impedance, hence an active probe with 1 MΩ input and 50 Ω output impedance was placed between AUT output pin and network analyzer input pin (see Fig. B.5) to prevent the AUT from driving a resistive load being extensively low.

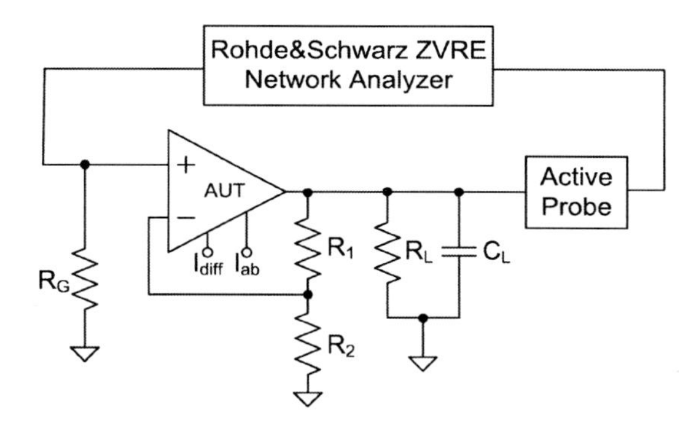

Figure B.5: Unity gain frequency and phase margin measurement setup.

B.5 Offset Voltage

The input-referred offset voltage has just been determined for the OP130b design using the setup of Fig. B.6.

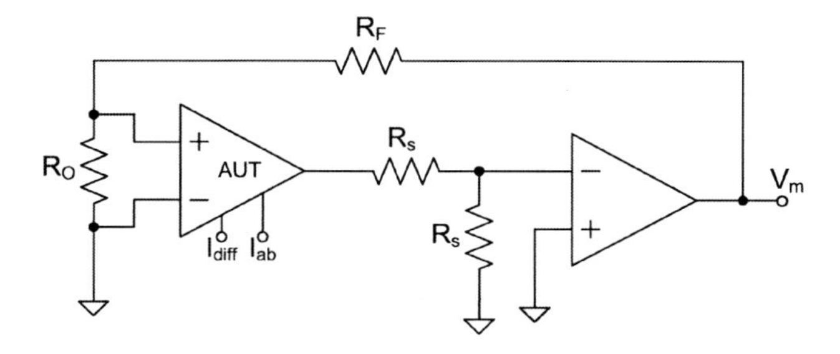

Figure B.6: Measurement of the input-referred offset voltage.

The circuit uses an auxiliary amplifier, similar to the open-loop gain measurement. Here V_s is set to signal ground and the input-referred offset voltage V_{os} is related to the measured voltage V_m by

$$V_{os} = V_m \frac{R_O}{R_F} \tag{B.4}$$

Resistor values of $R_F = 100\,\mathrm{k\Omega}$ and $R_O = 10\,\Omega$ were used for the OP130b measurement setup.

B.6 Total Harmonic Distortion

In order to determine the total harmonic distortion (THD) of the AUT, a 1 kHz, 1 Vpp single-tone test signal was applied to the AUT in a voltage follower configuration, see Fig. B.7. A Rohde & Schwarz UPL audio analyzer was used to generate the test signal and simultaneously measure the AUT output voltage. The THD was directly analyzed by the audio analyzer.

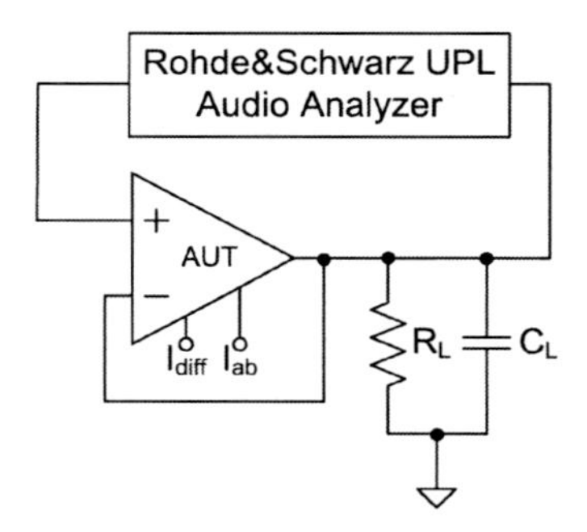

Figure B.7: Total harmonic distortion measurement

B.7 Slew Rate

The slew rate measurement setup shown in Fig. B.8 is comprised of function generator, the AUT in voltage follower configuration and an Agilent DSO3062A oscilloscope [120] which includes an automatic measurement mode to determine rise and fall times. The function generator was set to output a 1 Vpp rectangular pulse train and the resulting rise and fall times of the output signals were analyzed by the oscilloscope. The slew rate was calculated by relating these times to the output voltage change of 1 V.

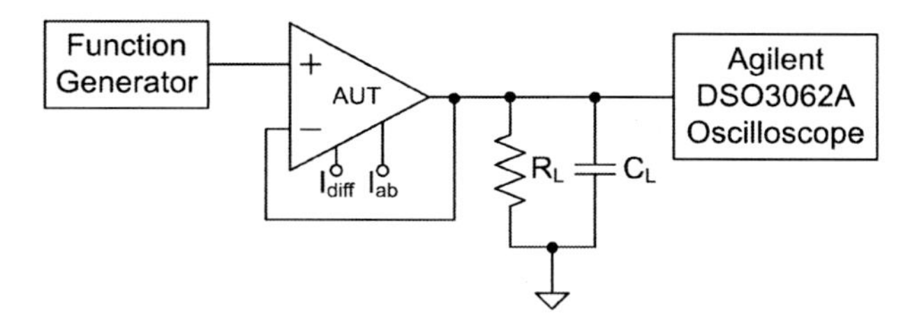

Figure B.8: Slew rate measurement

B.8 Common Mode Rejection Ratio

In order to circumvent resistor mismatch related problems in CMRR measurements an approach using an auxiliary op amp as depicted in Fig. B.9 was used for the OP130b. Its measurement principle is based on shifting the signal reference by varying V_{DD} and V_{SS} while keeping the overall supply voltage $|V_{DD}|+|V_{SS}|$ constant. In the first measurement, V_{DD} was set to +900 mV and V_{SS} to -300 mV with the original signal ground being now at +300 mV. Now, in order to drive the output voltage of the AUT to the original signal ground voltage, V_{s-} was set to -300 mV and the offset voltage V_m was measured (V_{m1}). The input voltage of the AUT, needed to drive V_{s+} to the original signal ground, is the attenuated V_m voltage as a result of the voltage division due to R_F and R_O. The same procedure was performed by setting V_{DD} to +300 mV, V_{SS} to -900 mV and V_{s-} to +300 mV. The voltage V_m was again measured (V_{m2}) and the CMRR was calculated using

$$\text{CMRR} = \left(\frac{R_O}{R_F + R_O}\right) \frac{V_{m2} - V_{m1}}{V_{s+} - V_{s-}} \tag{B.5}$$

The resistor values used were $R_F = 100\,\text{k}\Omega$, $R_O = 10\,\Omega$ and $R_S = 100\,\text{k}\Omega$.

B.9 Power Supply Rejection Ratio

The PSRR measurement uses the same test setup like the CMRR measurement in Fig. B.9. However, the measurement procedure differs in two aspects, first, the overall supply voltage is no longer constant and second, the voltage V_{s-} is held at signal ground. In order to determine the PSRR of the OP130b with respect to changes in the positive and negative supply voltage, V_{DD} and V_{SS} were first set to +400 mV and -600mV and the offset voltage V_m was measured (V_{m1}). This procedure was repeated by setting V_{DD} to +600 mV and V_{SS} to -400 mV and noting V_m as

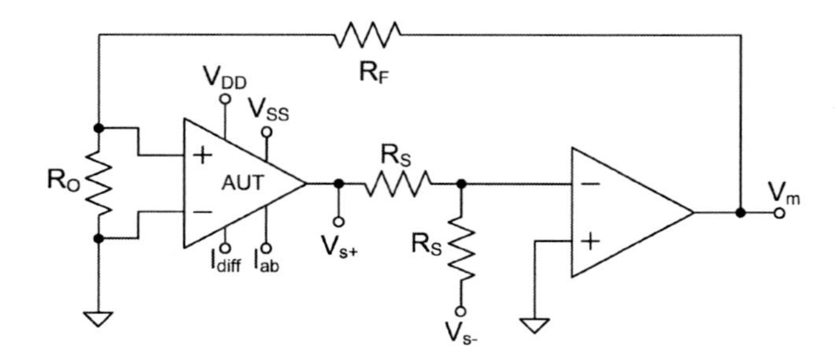

Figure B.9: CMRR and PSRR measurement setup

V_{m2}. The resulting V_{DD} and V_{SS} related PSRR values were calculated using

$$\text{PSRR}_+ = \left(\frac{R_O}{R_F + R_O}\right) \frac{V_{m2} - V_{m1}}{V_{DD}} \quad \text{(B.6)}$$

$$\text{PSRR}_- = \left(\frac{R_O}{R_F + R_O}\right) \frac{V_{m2} - V_{m1}}{V_{SS}} \quad \text{(B.7)}$$

B.10 Constant-gm Measurement

The performance of the constant-gm circuit was investigated using the setup shown in Fig. B.10.

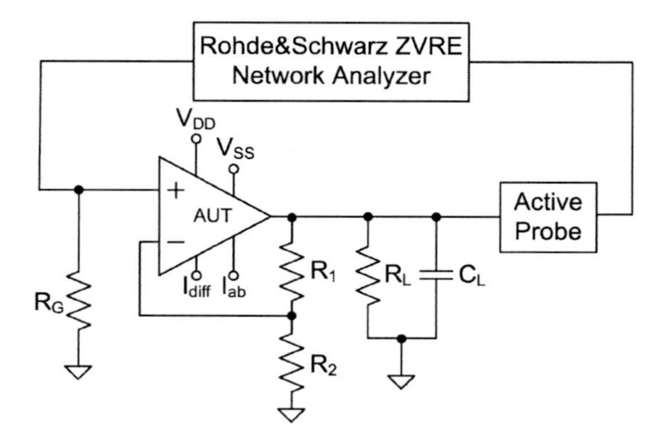

Figure B.10: Constant-gm measurement

It is essentially the same test circuit like for the unity gain measurement setup, however, now the signal ground is variable by changing V_{DD} and V_{SS} accordingly. The signal ground was stepwise changed within the supply voltage range. For every single step, the unity gain frequency f_u (which is related to the overall input transconductance of the AUT as described in Section 4.4.5) was determined.

List of Figures

List of Tables

List of Abbreviations and Symbols

Frequently used abbreviations and symbols (including units) are given below.

Abbreviations

SiO_2	Silicon dioxide
ADC	Analog-to-digital converter
AFE	Analog front-end
Ag/AgCl	Silver/Silver-chloride
AUT	Amplifier under test
BIST	Built-in self test
BSIM	Berkeley Short-channel IGFET Model
CMOS	Complementary metal-oxide-semiconductor
DSP	Digital signal processor
ECG	Electrocardiogram
EEG	Electroencephalogram
EMG	Electromyogram
ENOB	Effective number of bits
EP	Evoked Potentials
FOM	Figure of merit
HP	High-pass
IA	Instrumentation amplifier
IA130	130 nm IA design
LP	Low-pass
LSB	Least significant bit
M3C2	Three channel biomedical SoC
M3C3	Ten channel biomedical signal acquisition system
MOSFET	Metal-oxide-semiconductor field-effect transistor
NMOS	N-channel MOS
OP130a	130 nm programmable op amp designs with no chopper
OP130b	130 nm programmable op amp designs including chopper
OP350a/b	350 nm programmable op amp designs
PGA	Programmable gain amplifier

PMOS	P-channel MOS	
PSP	Postsynaptic potentials	
RMS	Root-mean-square	
RSCE	Reverse short channel effect	
RU	Respiration unit	
SCE	Short channel effect	
Si	Silicon	
SKLP	Sallen-Key low-pass filter	
SoC	System-on-chip	
SPICE	Simulation program with integrated circuit emphasis	
TCU	Timing and control unit	
UART	Universal asynchronous receiver/transmitter	

Symbols

β	Scaled current factor	(A/V^2)
$\Delta A_{12/3}$	First/second stage mismatch gain of the 3-op amp IA	(-)
ΔA_{2IA}	2-op amp IA gain including gain error	(-)
Δ_{2IA}	Relative error in resistor ratio of the 2-op amp IA	(-)
$\Delta_{3IA_{12/3}}$	Relative error in resistor ratio of the 3-op amp IA (first/second stage)	(-)
δ_{OP}	Current ratio I_q/I_{ab}	(-)
γ	Body effect parameter	($\sqrt{\text{V}}$)
γ_{OP}	Current ratio I_{FCS}/I_{ab}	(-)
κ_{PM}	Input to output stage transconductance ratio for stability	(-)
λ	Output resistance parameter	(1/V)
$(V_{GS})_{sv}$	Transition V_{GS} strong inversion/velocity saturation	(V)
$(V_{GS})_{ws}$	Transition V_{GS} weak inversion/strong inversion	(V)
CMRR	Common-mode rejection ratio	(-)
$\text{CMRR}_{\Delta OP}$	IA CMRR due to to finite op amp open-loop gain	(-)
$\text{CMRR}_{\Delta R}$	IA CMRR due to resistor mismatch	(-)
DR	Dynamic range	(-)
DSOR	DC-suppression to standard DC-offset range ratio	(-)
ICMR	Input common-mode range	(V)
k	Boltzmann constant	(J/K)
NEF	Noise efficiency factor	(-)
PSD	Power spectral density	(V^2/Hz)
PSRR	Power supply rejection ratio	(-)
q	Electron charge	(C)
SNR	Signal-to-noise ratio	(-)
THD	Total harmonic distortion	(-)

T	Temperature	(K)
μ_0	Low-field surface mobility	(m^2/Vs)
ν_{sat}	carrier saturation velocity	(m/s)
ω_c/f_c	Low-pass cut-off frequency	(Hz)
ω_T	Transit frequency	(Hz)
ω_u/f_u	Unity gain frequency	(Hz)
ω_p	Pole frequency	(Hz)
σ	Standard deviation	(-)
σ^2	Variance	(-)
A	Amplifier gain	(-)
AF	Drain current 1/f noise parameter	(-)
A_d	Differential-mode gain	(-)
A_β	Current factor mismatch constant	(% m)
$A_{\Delta 2IA}$	Mismatch gain of the 3-op amp IA	(-)
$A_{\Delta 3IA}$	3-op amp IA gain including gain error	(-)
$A_{cm,\Delta OP}$	IA common-mode gain due to finite op amp open-loop gain	(-)
$A_{cm,\Delta R}$	IA common-mode gain due to resistor mismatch	(-)
A_{cm}	Common-mode gain	(-)
A_{d0}	DC open loop gain	(-)
A_{v0}	Intrinsic gain	(-)
$A_{V_{TH}}$	Threshold voltage mismatch constant	(Vm^2)
BW	Bandwidth	(Hz)
C	Capacitance	(F)
C_A	First stage input capacitance	(F)
C_B	Second stage input capacitance	(F)
C_c	Miller capacitance	(F)
C_{db}	Drain-bulk capacitance	(F)
C'_d	Depletion capacitance per unit area	(F/m^2)
C_{gb}	Gate-bulk capacitance	(F)
C_{gd}	Gate-drain capacitance	(F)
C_{gs}	Gate-source capacitance	(F)
$C_{ov,d}$	Gate overlap capacitance at the drain	(F)
$C_{ov,s}$	Gate overlap capacitance at the source	(F)
C'_{ox}	Gate oxide capacitance per unit area	(F/m^2)
C_{sb}	Source-bulk capacitance	(F)
E_{sat}	Critical field for velocity saturation	(V/m)
f	Frequency	(Hz)
f_k	1/f noise corner frequency	(Hz)
f_{ch}	Chopper clock frequency	(Hz)

g_m	Transconductance	(S)
g_{d0}	Channel resistance at zero drain-source bias	(Ω)
g_{ds}	Output conductance	(S)
g_{mA}	First stage transconductance	(S)
g_{mB}	Second stage transconductance	(S)
g_{mb}	bulk transconductance	(S)
I	Current	(A)
i	Small-signal current	(A)
I_D	Drain current	(A)
I_{ab}	Op amp output stage (programming) current	(A)
I_{bias}	M3C2/M3C3 front-end bias current	(A)
I_{D0}	Normalized specific current	(A)
I_{diff}	Op amp input stage (programming) current	(A)
I_{FCS}	Floating current source current	(A)
$i^2_{nf,I_D}(f)$	Drain current 1/f noise PSD	(A^2/Hz)
i_{no}	Output RMS current noise	(A)
$i^2_{nth,I_D}(f)$	Drain current thermal noise PSD	(A^2/Hz)
i_n	RMS current noise	(A)
$i_n(f)$	Current noise spectral density	($A/\sqrt{Hz}$)
i^2_n	Total current noise power	(A^2/Hz)
$i^2_n(f)$	Power spectral density of a current noise	(A^2/Hz)
KF	1/f noise parameter	(A/V)
KF'	Scaled 1/f noise parameter	(C^2/m^2)
KP_n	Current factor	(A/V^2)
k_s	MOSFET scaling factor	(-)
L	Channel length	(m)
L_{min}	Minimal channel length	(m)
n	Factor for body effect and subthreshold slope	(-)
P	Power	(W)
PM	Phase margin	(°)
$P_{low/med/high}$	Programmable power dissipation	(W)
R	Resistance	(Ω)
r_A	First stage output resistance	(Ω)
R_L	Load resistor value	(Ω)
r_{ds}	Output resistance	(Ω)
t	Time	(s)
U_T	Thermal voltage	(V)
V	Voltage	(V)
v	Small-signal voltage	(V)

V_A	MOSFET "Early voltage"	(V)
V_A'	MOSFET-"Early voltage" approximation	(V)
V_{bcn}	Cascode bias voltage - NMOS part	(V)
V_{bcp}	Cascode bias voltage - PMOS part	(V)
V_{DD}	Supply voltage	(V)
V_{DSAT}	Saturation voltage	(V)
V_{DS}	Source-bulk voltage	(V)
V_{GS}	Gate-source voltage	(V)
V_{i+}	Non-inverting input voltage	(V)
$V_{i,c}$	Common-mode input voltage	(V)
$V_{i,d}$	Differential input voltage	(V)
V_{i-}	Inverting input voltage	(V)
V_i	Input voltage	(V)
V_{LSB}	Least significant bit voltage	(V)
$v^2_{nif,I_D}(f)$	MOSFET 1/f voltage noise PSD	(V^2/Hz)
$v^2_{nif,OP}(f)$	Input-referred 1/f voltage noise PSD - op amp	(V^2/Hz)
$v^2_{nith,I_D}(f)$	MOSFET thermal voltage noise PSD	(V^2/Hz)
$v^2_{nith,OP}(f)$	Input-referred thermal voltage noise PSD- op amp	(V^2/Hz)
v_{ni}	Input-referred RMS voltage noise	(V)
v_{no}	Output RMS voltage noise	(V)
v_n	RMS voltage noise	(V)
$v_n(f)$	Voltage noise spectral density	($V/\sqrt{Hz}$)
v_n^2	Total voltage noise power	(V^2/Hz)
$v_n^2(f)$	Power spectral density of a voltage noise	(V^2/Hz)
V_{OV}	Overdrive voltage	(V)
V_o	Output voltage	(V)
V_{SB}	Source-bulk voltage	(V)
V_{sgnd}	Signal ground voltage	(V)
V_{SS}	Negative supply voltage/Ground	(V)
V_{T0}	Threshold voltage for $V_{BS} = 0$	(V)
V_{THn}	NMOS threshold voltage	(V)
V_{THp}	PMOS threshold voltage	(V)
V_{TH}	Threshold voltage	(V)
W	Channel width	(m)
$V_{DSAT}{}^*$	Long channel V_{DSAT} (= V_{OV})	(V)
$V_{DC,set}$	DC-suppression control voltage	(V)

Bibliography

[1] Google Scholar, Google Inc., Mountain View, USA. Search results retrieved September 2009. [Online]. Available: http://scholar.google.com/

[2] Federal Statistical Office of Germany, *Im Jahr 2050 doppelt so viele 60-Jährige wie Neugeborene*, Press release No. 464, 7. Sept. 2006.

[3] "Intel Developer Forum Manufacturing Keynote Disclosuress," Press release, Intel Developer Forum, San Francisco, 22.-24. September 2009.

[4] S. H. Wright, "Generation of resting membrane potential." *Adv Physiol Educ*, vol. 28, no. 1-4, pp. 139–142, Dec 2004.

[5] J. G. Webster, Ed., *Bioinstrumentation.* New York: John Wiley & Sons, 2004.

[6] J. J. Carr and J. M. Brown, *Introduction to Biomedical Equipment Technology.* Upper Saddle River: Prentice-Hall, 2001.

[7] R. Plonsey and R. C. Barr, *Bioelectricity: A Quantitative Approach*, 3rd ed. New York: Springer, 2007.

[8] S. Institute, "BioFEM: A SCIRun PowerApp that computes the electric field in a volume produced by a set of dipoles. Scientific Computing and Imaging Institute (SCI)," 2009. [Online]. Available: http://www.sci.utah.edu/cibc/software/index.html

[9] (2007, January) How the heart works. National Heart, Lung, and Blood Institute (NHLBI). [Online]. Available: http://www.nhlbi.nih.gov/health/dci/Diseases/hhw/hhw_whatis.html

[10] S. Zschocke, *Klinische Elektroenzephalographie*, 2nd ed. Heidelberg: Springer, 2002.

[11] K. Blinowska and P. Durka, *Wiley Encyclopedia Of Biomedical Engineering.* John Wiley & Sons, 2006, vol. 2, ch. Electroencephalography (EEG), pp. 1341–1355.

[12] L. Abu-Saleh, J. M. Tomasik, W. Galjan, and W. H. Krautschneider, "A SoC based mobile EEG signal acquisition system using multi-sensor-recording to reduce noise and artifacts," in *Proc. ProRISC 2008*, Nov. 27.-28. 2008, pp. 291–294.

[13] J. G. Webster, *Medical Instrumentation: Application and Design*, 3rd ed. New York: John Wiley & Sons, 1998.

[14] J. D. Bronzino, Ed., *The Biomedical Engineering Handbook*, 2nd ed. Boca Raton: CRC Press, 2000.

[15] Nuwer *et al.*, "IFCN standards for digital recording of clinical EEG," *Electroencephalography and Clinical Neurophysiology*, vol. 106, Issue 3, pp. 259–261, March 1998.

[16] C. J. Harland, T. Clark, and R. J. Prance, "Electric potential probes-new directions in the remote sensing of the human body," *Measurement Science and Technology*, vol. 13, pp. 163–169, 2002.

[17] A. Ueno, Y. Akabane, T. Kato, H. Hoshino, S. Kataoka, and Y. Ishiyama, "Capacitive sensing of electrocardiographic potential through cloth from the dorsal surface of the body in a supine position: A preliminary study," *IEEE Trans. Biomed. Eng.*, vol. 54, no. 4, pp. 759–766, 2007.

[18] R. A. Normann, "Microfabricated electrode arrays for restoring lost sensory and motor functions," in *Proc. TRANSDUCERS, Solid-State Sensors, Actuators and Microsystems, 12th International Conference on*, vol. 2, 2003, pp. 959–962.

[19] L. A. Geddes and L. E. Baker, *Principles of Applied Biomedical Instrumentation*, 3rd ed. New York: John Wiley & Sons, 1989.

[20] *Disposable ECG electrodes*, ANSI/AAMI Std. EC12:2000/(R)2005.

[21] H. de Talhouet and J. G. Webster, "The origin of skin-stretch-caused motion artifacts under electrodes." *Physiol Meas*, vol. 17, no. 2, pp. 81–93, May 1996.

[22] B. R. Eggins, "Skin contact electrodes for medical applications." *Analyst*, vol. 118, no. 4, pp. 439–442, Apr 1993.

[23] A. Bolz and W. Urbaszek, *Technik in der Kardiologie.* Berlin: Springer, 2002.

[24] H. W. Tam and J. G. Webster, "Minimizing electrode motion artifact by skin abrasion," *IEEE Trans. Biomed. Eng.*, vol. BME-24, no. 2, pp. 134–139, 1977.

[25] B. Fuchs, "Integrierte Sensorschaltungen zur EKG- und EEG-Ableitung mit prädiktiver Signalverarbeitung," Ph.D. dissertation, Hamburg University of Technology, Shaker Verlag Aachen, 2004.

[26] J. Moore and G. Zouridakis, Eds., *Biomedical Technology and Devices Handbook.* Boca Raton: CRC Press, 2004.

[27] M. S. J. Steyaert, W. M. C. Sansen, and C. Zhongyuan, "A micropower lownoise monolithic instrumentation amplifier for medical purposes," *IEEE J. Solid-State Circuits*, vol. 22, no. 6, pp. 1163–1168, Dec 1987.

[28] *Diagnostic Electrocardiographic Devices*, ANSI/AAMI Std. EC11:1991.

[29] E. Spinelli, N. Martinez, M. A. Mayosky, and R. Pallas-Areny, "A novel fully differential biopotential amplifier with DC suppression," *IEEE Trans. Biomed. Eng.*, vol. 51, no. 8, pp. 1444–1448, Aug. 2004.

[30] C. Bronskowski, "Programmierbarer Operationsverstärker zur Erfassung von bioelektrischen Signalen mit unterschiedlichen Rausch- und Verlustleistungs-Anforderungen," Ph.D. dissertation, Hamburg University of Technology, Shaker Verlag Aachen, 2006.

[31] R. R. Harrison, "The design of integrated circuits to observe brain activity," *Proceedings of the IEEE*, vol. 96, no. 7, pp. 1203–1216, July 2007.

[32] R. F. Yazicioglu, P. Merken, R. Puers, and C. Van Hoof, "A 200 μw eight-channel EEG acquisition ASIC for ambulatory EEG systems," *IEEE J. Solid-State Circuits*, vol. 43, no. 12, pp. 3025–3038, Dec. 2008.

[33] R. H. Dennard, F. H. Gaensslen, V. L. Rideout, V. L. , E. Bassous, and A. R. LeBlanc, "Design of ion-implanted MOSFET's with very small physical dimensions," *IEEE J. Solid-State Circuits*, vol. 9, no. 5, pp. 256–268, Oct 1974.

[34] M. Bohr, "A 30 year retrospective on Dennard's MOSFET scaling paper," *IEEE SSCS NEWSLETTER*, vol. 12, no. 1, Winter 2007.

[35] A. J. Annema, B. Nauta, R. van Langevelde, and H. Tuinhout, "Analog circuits in ultra-deep-submicron CMOS," *IEEE J. Solid-State Circuits*, vol. 40, no. 1, pp. 132–143, Jan. 2005.

[36] K. Bult, "Analog design in deep sub-micron CMOS," in *Proc. IEEE ESSCIRC 2000*, 19–21 Sept. 2000, pp. 126–132.

[37] M. Vertregt, "The analog challenge of nanometer CMOS," in *Proc. International Electron Devices Meeting IEDM '06*, 11–13 Dec. 2006, pp. 1–8.

[38] J. H. Huijsing, K.-J. de Langen, R. Hogervorst, and R. G. H. Eschauzier, "Low-voltage low-power opamp based amplifiers," *Analog Integrated Circuits and Signal Processing*, vol. 8, no. 1, pp. 49–67, July 1995.

[39] J. H. Huijsing, *Operational Amplifiers - Theory and Design*. Dordrecht, The Netherlands: Kluwer Academic Publishers, 2001.

[40] EECS Department, University of California, Berkeley, "SPICE (Simulation Program with Integrated Circuit Emphasis) and BSIM (Berkeley Short-channel IGFET Model)." [Online]. Available: http://www.eecs.berkeley.edu/

[41] F. Silveira, D. Flandre, and P. G. A. Jespers, "A gm/ID based methodology for the design of CMOS analog circuits and its application to the synthesis of a silicon-on-insulator micropower OTA," *IEEE J. Solid-State Circuits*, vol. 31, no. 9, pp. 1314–1319, Sept. 1996.

[42] Y. Tsividis, *Operation and Modeling of the MOS Transistor*, 2nd ed. Boston: McGraw-Hill, 1999.

[43] N. G. Einspruch and G. S. Gildenblatt, Eds., *Advanced MOS Device Physics*, ser. VLSI Electronics: Microstructure Science. San Diego: Academic Press, 1989, vol. 18.

[44] W. Liu et al., *BSIM3v3.3 MOSFET Model User's Manual*, University of California, Department of Electrical Engineering and Computer Sciences, Berkeley, CA, 2005.

[45] W. M. C. Sansen, *Analog Design Essentials*. Dordrecht, The Netherlands: Springer, 2006.

[46] J. H. Huang, Z. H. Liu, M. C. Jeng, P. K. Ko, and C. Hu, "A physical model for MOSFET output resistance," in *Technical Digest. International Electron Devices Meeting (IEDM) 1992*, Dec. 13.-16. 1992, pp. 569–572.

[47] K. R. Laker and W. M. C. Sansen, *Design of Analog Integrated Circuits and Systems*. New York: McGraw-Hill, 1994.

[48] E. Vittoz and J. Fellrath, "CMOS analog integrated circuits based on weak inversion operations," *IEEE J. Solid-State Circuits*, vol. 12, no. 3, pp. 224–231, Jun 1977.

[49] C. C. Enz, F. Krummenacher, and E. A. Vittoz, "An analytical MOS transistor model valid in all regions of operation and dedicated to low-voltage and low-current applications," *J. Analog Int. Circ. Signal Processing*, vol. 8, no. 1, pp. 83–114, July 1995.

[50] M. Lundstrom, "ECE 612: Nanoscale Transistors (Fall 2008)," Aug 2008. [Online]. Available: http://nanohub.org/resources/5328

[51] P. R. Gray, P. J. Hurst, S. H. Lewis, and R. G. Meyer, *Analysis and Design of Analog Integrated Circuits*, 4th ed. New York: John Wiley & Sons, 2001.

[52] B. Razavi, *Design of Analog CMOS Integrated Circuits.* Boston: McGraw-Hill, 2001.

[53] R. J. Baker, *CMOS: Circuit Design, Layout, and Simulation*, 2nd ed. New York: John Wiley & Sons, 2007.

[54] C. D. Motchenbacher and J. A. Connelly, *Low-Noise Electronic System Design.* New York: John Wiley & Sons, 1993.

[55] A. van der Ziel, *Noise in Solid State Devices and Circuits.* New York: John Wiley & Sons, 1986.

[56] R. P. Jindal, "Compact noise models for MOSFETs," *IEEE Trans. Electron Devices*, vol. 53, no. 9, pp. 2051–2061, Sept. 2006.

[57] A. J. Scholten, L. F., R. van Langevelde, R. J. Havens, A. T. A. Zegers-van Duijnhoven, and V. C. Venezia, "Noise modeling for RF CMOS circuit simulation," *IEEE Trans. Electron Devices*, vol. 50, no. 3, pp. 618–632, March 2003.

[58] A. L. McWhorter, *Semiconductor Surface Physics.* Philadelphia, PA: Univ. of Pennsylvania Press, 1957, ch. 1/f noise and germanium surface properties, pp. 207–228.

[59] M. J. Uren, D. J. Day, and M. J. Kirton, "1/f and random telegraph noise in silicon metal-oxide-semiconductor field-effect transistors," *Appl. Phys. Lett.*, vol. 47, no. 11, pp. 1195–1197, Dec. 1985.

[60] F. N. Hooge, "1/f noise is no surface effect," *Physics Letters A*, vol. 29, no. 3, pp. 139–140, 1969.

[61] F. N. Hooge and L. K. J. Vandamme, "Lattice scattering causes 1/f noise," *Physics Letters A*, vol. 66, no. 4, pp. 315–316, 1978.

[62] E. P. Vandamme and L. K. J. Vandamme, "Critical discussion on unified 1/f noise models for MOSFETs," *IEEE Trans. Electron Devices*, vol. 47, no. 11, pp. 2146–2152, Nov. 2000.

[63] R. Jayaraman and C. G. Sodini, "A 1/noise technique to extract the oxide trap density near the conduction band edge of silicon," *IEEE Trans. Electron Devices*, vol. 36, no. 9, pp. 1773–1782, Sept. 1989.

[64] K. K. Hung, P. K. Ko, C. Hu, and Y. C. Cheng, "A unified model for the flicker noise in metal-oxide-semiconductor field-effect transistors," *IEEE Trans. Electron Devices*, vol. 37, no. 3, pp. 654–665, March 1990.

[65] M. J. M. Pelgrom, A. C. J. Duinmaijer, and A. P. G. Welbers, "Matching properties of MOS transistors," *IEEE J. Solid-State Circuits*, vol. 24, no. 5, pp. 1433–1439, Oct 1989.

[66] P. R. Kinget, "Device mismatch and tradeoffs in the design of analog circuits," *IEEE J. Solid-State Circuits*, vol. 40, pp. 1212–1224, June 2005.

[67] S. Thompson, P. Packan, and M. Bohr, "MOS scaling: Transistor challenges for the 21st century," *Intel Technol. J.*, vol. Q3, pp. 1–19, 1998.

[68] N. S. Kim, T. Austin, D. Baauw, T. Mudge, K. Flautner, J. S. Hu, M. J. Irwin, M. Kandemir, and V. Narayanan, "Leakage current: Moore's law meets static power," *Computer*, vol. 36, no. 12, pp. 68–75, Dec. 2003.

[69] G. Gielen and W. Dehaene, "Analog and digital circuit design in 65 nm CMOS: end of the road?" in *Proc. Design, Automation and Test in Europe*, 2005, pp. 37–42.

[70] F. Schlögl, K. Schneider-Hornstein, and H. Zimmermann, "Gain reduction by gate-leakage currents in regulated cascodes," in *Proc. 11th IEEE Workshop on Design and Diagnostics of Electronic Circuits and Systems DDECS 2008*, 16-18 April 2008, pp. 1–4.

[71] Weixun Yan, R. Kolm, and H. Zimmermann, "A low-voltage low-power fully differential rail-to-rail input/output opamp in 65-nm CMOS," in *Proc. IEEE*

International Symposium on Circuits and Systems ISCAS 2008, 18–21 May 2008, pp. 2274–2277.

[72] K. Mistry *et al.*, "A 45nm Logic Technology with High-k+Metal Gate Transistors, Strained Silicon, 9 Cu Interconnect Layers, 193nm Dry Patterning, and 100% Pb-free Packaging," in *Proc. IEEE International Electron Devices Meeting IEDM 2007*, 10-12 Dec. 2007, pp. 247–250.

[73] C. H. Wann, K. Noda, T. Tanaka, M. Yoshida, and Chenming Huang, "A comparative study of advanced MOSFET concepts," *IEEE Trans. Electron Devices*, vol. 43, no. 10, pp. 1742–1753, Oct. 1996.

[74] J. Szynowski., "CMRR analysis of instrumentation amplifiers," *Electronics Letters*, vol. 19, no. 14, pp. 547–549, July 7 1983.

[75] J. Karki, "Analysis of the Sallen-Key architecture (rev. b)," Texas Instruments Application Report, Sep. 2002, SLOA024B.

[76] J. A. Connelly, Ed., *Analog Integrated Circuits.* New York: John Wiley & Sons, 1975.

[77] R. Hogervorst, J. P. Tero, R. G. H. Eschauzier, and J. H. Huijsing, "A compact power-efficient 3 V CMOS rail-to-rail input/output operational amplifier for VLSI cell libraries," *IEEE J. Solid-State Circuits*, vol. 29, no. 12, pp. 1505–1513, Dec. 1994.

[78] R. Hogervorst and J. H. Huijsing, *Design of Low-voltage, Low-power Operational Amplifier Cells.* Dordrecht, The Netherlands: Kluwer Academic Publishers, 1996.

[79] A. S. Sedra and K. C. Smith, *Microelectronic Circuits.* Oxford: Oxford University Press, 2004.

[80] F. You, S. H. K.Embabi, and E. Sanchez-Sinencio, "On the common mode rejection ratio in low voltage operational amplifiers with complementary N-P input pairs," *IEEE Trans. Circuits Syst. II, Analog Digit. Signal Process.*, vol. 44, no. 8, pp. 678–683, Aug. 1997.

[81] P. Meier auf der Heide, C. Bronskowski, J. M. Tomasik, and D. Schroeder, "A CMOS operational amplifier with constant 68° phase margin over its whole range of noise-power trade-off programmability," in *Proc. IEEE ESSCIRC 2007*, Munich, Germany, Sep. 11.-13. 2007, pp. 452–455.

[82] C. Bronskowski and D. Schroeder, "An ultra low-noise CMOS operational amplifier with programmable noise-power trade-off," in *Proc. ESSCIRC 2006*, Grenoble, France, 19–21 Sept. 2006, pp. 368–371.

[83] ——, "Systematic design of programmable operational amplifiers with noise-power trade-off," *IET Circuits, Devices & Systems*, vol. 1, no. 1, pp. 41–48, Feb. 2007.

[84] J. M. Tomasik, K. M. Hafkemeyer, W. Galjan, D. Schroeder, and W. H. Krautschneider, "A 130nm CMOS programmable operational amplifier," in *Proc. IEEE Norchip 2008*, Tallin, Estonia, Nov. 17.-18. 2008, pp. 29–32.

[85] F. Schlögl and H. Zimmermann, "120nm CMOS OPAMP with 690 MHz f_t and 128 dB DC gain," in *Proc. IEEE ESSCIRC 2005*, Grenoble, France, 19.-21. Sept. 2005, pp. 251–254.

[86] Weixun Yan and H. Zimmermann, "A 120nm CMOS fully differential rail-to-rail I/O opamp with highly constant signal behavior," in *Proc. International SOC Conference 2006*, Sept. 2006, pp. 3–6.

[87] J. M. Tomasik, D. Schroeder, and W. Krautschneider, "Analog-design mit nanometer-MOSFETs," Workshop *Integrierte Analogschaltungen*, Berlin, Germany, Mar. 10.-11. 2008, (Oral presentation).

[88] C. C. Enz and G. C. Temes, "Circuit techniques for reducing the effects of op-amp imperfections: autozeroing, correlated double sampling, and chopper stabilization," *Proceedings of the IEEE*, vol. 84, no. 11, pp. 1584–1614, Nov. 1996.

[89] J. M. Tomasik, C. Bronskowski, and W. H. Krautschneider, "A model for switched biasing MOSFET 1/f noise reduction," in *Proc. SAFE 2005*, Veldhoven, The Netherlands, Nov. 17.-18. 2005, pp. 60–63.

[90] J. M. Tomasik and W. H. Krautschneider, "Review of 1/f noise in MOSFETs and development of an experimental opamp using switched biasing to reduce 1/f noise," Kleinheubacher Tagung 2006, Miltenberg, Germany, Sep. 25.-29. 2006, (Oral presentation).

[91] R. Rieger and J. T. Taylor, "Design strategies for multi-channel low-noise recording systems," *J. Analog Int. Circ. Signal Processing*, vol. 58, no. 2, pp. 123–133, Feb. 2009.

[92] Project MyoPlant, funded by the German Ministry of Education and Research (BMBF) (FKZ 16SV3699).

[93] Project M3C, funded by the European Commision, DG RTD, Contract No. COOP-CT-2004-508291-M3C.

[94] N. V. Helleputte, A. Mora-Sanchez, W. Galjan, J. M. Tomasik, D. Schroeder, W. H. Krautschneider, and R. Puers, "A flexible system-on-chip (SoC) for biomedical signal acquisition and processing," in *Proc. EUROSENSORS XX*, Goeteborg, Sweden, Sep. 17.-20. 2006, pp. 288–289.

[95] J. Tomasik and W. Galjan, "M3C: A Multi Monitoring Medical Chip for Homecare Applications," in Activity Report 2006, Europractice IC Service, 2007.

[96] N. V. Helleputte, J. M. Tomasik, W. Galjan, A. Mora-Sanchez, D. Schroeder, W. H. Krautschneider, and R. Puers, "A flexible system-on-chip (SoC) for biomedical signal acquisition and processing," *Sensors and Actuators A: Physical*, vol. 142, Issue 1, pp. 361–368, March 2008.

[97] A. Mora-Sanchez, D. Schroeder, and W. H. Krautschneider, "Sigma-delta modulators of 2nd- and 3rd-order with a single operational transconductance amplifier for low-power analogue-to-digital conversion," in *Proc. ProRisc 2005*, Veldhoven, The Netherlands, Nov. 17.-18. 2005, pp. 259–262.

[98] A. Mora-Sanchez and D. Schroeder, "Low-power decimation filter in a 0.35 μm CMOS technology for a multi-channel biomedical data acquisition chip," in *Proc. XI Iberchip Workshop*, Salvador do Bahia, Brasil, 2005, pp. 199–202.

[99] Evatronix SA, Bielsko Biala, Poland, http://www.evatronix.pl.

[100] W. Galjan, D. Naydenova, J. M. Tomasik, D. Schroeder, and W. H. Krautschneider, "A portable SoC-based ECG-system for 24h x 7d operating time," in *Proc. IEEE Biocas 2008*, Baltimore, USA, Nov. 20.-22. 2008, pp. 85–88.

[101] A. K. Lu, G. W. Roberts, and D. A. Johns, "A high-quality analog oscillator using oversampling D/A conversion techniques," *IEEE Trans. Circuits Syst. II, Analog Digit. Signal Process.*, vol. 41, no. 7, pp. 437–444, July 1994.

[102] K. M. Hafkemeyer, W. Galjan, J. M. Tomasik, D. Schroeder, and W. Krautschneider, "System-on-chip approach for biomedical signal acquisition," in *Proc. ProRISC 2007*, Veldhoven, The Netherlands, Nov. 29.-30. 2007, pp. 26–29.

[103] W. Galjan, K. M. Hafkemeyer, J. M. Tomasik, F. Wagner, W. H. Krautschneider, and D. Schroeder, "Highly sensitive biomedical amplifier with CMRR calibration and DC-offset compensation," in *Proc. of IEEE EUROCON 2009*, Saint Petersburg, Russia, May 18.-23. 2009, pp. 172–175.

[104] F. Medeiro, B. Perez-Verdu, J. M. de la Rosa, and A. Rodriguez-Vazquez, "Using CAD tools for shortening the design cycle of high-performance ΣΔM: A 16.4bit 9.6kHz 1.71mW ΣΔM in CMOS 0.7μm technology," *International Journal of Circuit Theory and Applications*, vol. 25, pp. 319–334, 1997.

[105] P. E. Allen and D. R. Holberg, *CMOS Analog Circuit Design*, 2nd ed. New York: Oxford University Press, 2002.

[106] F. Wagner, C. Jakobi, J. M. Tomasik, K. M. Hafkemeyer, W. Galjan, D. Schroeder, and W. H. Krautschneider, "Design and implementation of an automated test environment for signal-acquisition ASICs," in *Proc. SCD 2008 Conference*, Dresden, Germany, Apr. 23.-24. 2008.

[107] F. Gerfers, C. Hack, M. Ortmanns, and Y. Manoli, "A 1.2 V, 200 μW rail-to-rail op amp with 90 THD using replica gain enhancement," in *Proc. IEEE ESSCIRC 2002*, Florence, Italy, Sept. 24.-26. 2002, pp. 175–178.

[108] F. Schlögl and H. Zimmermann, "Low-voltage operational amplifier in 0.12 μm digital CMOS technology," *IEE Proceedings Circuits, Devices and Systems*, vol. 151, no. 5, pp. 395–398, Oct. 2004.

[109] J. Citakovic, I. R. Nielsen, J. H. N. P. Asbeck, and P. Andreani, "A 0.8V, 7 μA, rail-to-rail input/output, constant g_m operational amplifier in standard digital 0.18μm CMOS," in *Proc. IEEE Norchip 2005*, Oulu, Finland, 21-22 Nov. 2005, pp. 54–57.

[110] Y. Haga and I. Kale, "Achieving rail-to-rail input operation using level-shift multiplexing technique for all CMOS op-amps," in *Proc. 51st IEEE Midwest Symposium on Circuits and Systems, MWSCAS 2008.*, Knoxville, USA, Aug. 10.-13. 2008, pp. 698–701.

[111] J. H. Nielsen and T. Lehmann, "An implantable CMOS amplifier for nerve signals," in *Proc. 8th IEEE International Conference on Electronics, Circuits and Systems ICECS 2001*, vol. 3, 2.-5. Sept. 2001, pp. 1183–1186.

[112] W. Dabrowski, P. Grybos, and A. M. Litke, "A low noise multichannel integrated circuit for recording neuronal signals using microelectrode arrays," *Biosensors and Bioelectronic*, vol. 19, no. 7, pp. 749–761, Feb. 2004.

[113] J. Sacristan and M. T. Oses, "Low noise amplifier for recording eng signals in implantable systems," in *Proc. International Symposium on Circuits and Systems ISCAS '04*, vol. 4, 23.-26. May 2004, pp. 33–36.

[114] T. Denison, K. Consoer, W. Santa, A.-T. Avestruz, J. Cooley, and A. Kelly, "A 2 μW 100 nV/$\sqrt{}$/hz chopper-stabilized instrumentation amplifier for chronic measurement of neural field potentials," *IEEE J. Solid-State Circuits*, vol. 42, no. 12, pp. 2934–2945, Dec. 2007.

[115] J. Ji and K. D. Wise, "An implantable CMOS circuit interface for multiplexed microelectrode recording arrays," *IEEE J. Solid-State Circuits*, vol. 27, no. 3, pp. 433–443, March 1992.

[116] P. Mohseni and K. Najafi, "A fully integrated neural recording amplifier with dc input stabilization," *IEEE Trans. Biomed. Eng.*, vol. 51, no. 5, pp. 832–837, May 2004.

[117] Intersil, "Recommended test procedures for operation amplifiers," Intersil Application Note, Nov. 1996, AN551.1.

[118] Keithley Instruments Inc., Cleveland, USA, http://www.keithley.com.

[119] Rohde & Schwarz GmbH & Co. KG, München, Germany, http://www.rohde-schwarz.de.

[120] Agilent Inc., Santa Clara, USA, http://www.agilent.com.

List of Publications

1. **J. M. Tomasik**, C. Bronskowski, and W. H. Krautschneider, "A model for switched biasing MOSFET 1/f noise reduction," in *Proc. SAFE 2005*, Veldhoven, The Netherlands, Nov. 17.-18. 2005, pp. 60–63.

2. N. Van Helleputte, A. Mora-Sanchez, W. Galjan, **J. M. Tomasik**, D. Schroeder, W. H. Krautschneider, and R. Puers, "A flexible system-on-chip (SoC) for biomedical signal acquisition and processing," in *Proc. EUROSENSORS XX*, Goeteborg, Sweden, Sep. 17.-20. 2006, pp. 288–289.

3. **J. M. Tomasik** and W. H. Krautschneider, "Review of 1/f noise in MOSFETs and development of an experimental opamp using switched biasing to reduce 1/f noise," Kleinheubacher Tagung 2006, Miltenberg, Germany, Sep. 25.-29. 2006, (Oral presentation).

4. P. Meier auf der Heide, C. Bronskowski, **J. M. Tomasik**, and D. Schroeder, "A CMOS operational amplifier with constant 68° phase margin over its whole range of noise-power trade-off programmability," in *Proc. IEEE ESSCIRC 2007*, Munich, Germany, Sep. 11.-13. 2007, pp. 452–455.

5. K. H. Hafkemeyer, W. Galjan, **J. M. Tomasik**, D. Schroeder, and W. H. Krautschneider, "System-on-chip approach for biomedical signal acquisition," in *Proc. ProRISC 2007*, Veldhoven, The Netherlands, Nov. 29.-30. 2007, pp. 26–29.

6. **J. M. Tomasik**, D. Schroeder, and W. H. Krautschneider, "Analog-design mit nanometer-MOSFETs," Workshop *Integrierte Analogschaltungen*, Berlin, Germany, Mar. 10.-11. 2008, (Oral presentation).

7. F. Wagner, C. Jakobi, **J. M. Tomasik**, K. M. Hafkemeyer, W. Galjan, D. Schroeder, and W. H. Krautschneider, "Design and implementation of an automated test environment for signal-acquisition ASICs," in *Proc. SCD 2008 Conference*, Dresden, Germany, Apr. 23.-24. 2008.

8. N. V. Helleputte, **J. M. Tomasik**, W. Galjan, A. Mora-Sanchez, D. Schroeder, W. H. Krautschneider, and R. Puers, "A flexible system-on-chip (SoC) for biomedical signal acquisition and processing," *Sensors and Actuators A: Physical*, vol. 142, Issue 1, pp. 361–368, March 2008.

9. L. Abu-Saleh, **J. M. Tomasik**, W. Galjan, and W. H. Krautschneider, "A SoC based mobile EEG signal acquisition system using multi-sensor-recording to reduce noise and artifacts," in *Proc. ProRISC 2008*, Nov. 27.-28. 2008, pp. 291–294.

10. **J. M. Tomasik**, K. M. Hafkemeyer, W. Galjan, D. Schroeder, and W. H. Krautschneider, "A 130nm CMOS programmable operational amplifier," in in Proc. IEEE Norchip 2008, Tallin, Estonia, Nov. 17.-18. 2008, pp. 29–32.

11. W. Galjan, D. Naydenova, **J. M. Tomasik**, D. Schroeder, and W. H. Krautschneider, "A portable SoC-based ECG-system for 24h x 7d operating time," in *Proc. IEEE Biocas 2008*, Baltimore, USA, Nov. 20.-22. 2008, pp. 85–88.

12. W. Galjan, K. M. Hafkemeyer, **J. M. Tomasik**, F. Wagner, W. H. Krautschneider, and D. Schroeder, "Highly sensitive biomedical amplifier with CMRR calibration and DC-offset compensation," in *Proc. of IEEE EUROCON 2009*, Saint Petersburg, Russia, May 18.-23. 2009, pp. 172–175.

Lebenslauf

Persönliche Angaben

Name:	Jakob Tomasik
Geburtsdatum:	13. August 1975
Geburtsort:	Gliwice (Polen)

Schule und Studium

Dezember 1998	Abitur, Abendgymnasium Hermaneum, Hamburg
1999 - 2004	Studium an der Technischen Universität Hamburg-Harburg, Studienrichtung: Informatikingenieurwesen
November 2004	Diplom-Ingenieur

Berufstätigkeit

05/1999 - 06/1999	Grundpraktikum bei der Deutschen Shell AG, Hamburg
08/2003 - 01/2004	Fachpraktikum bei der Philips Semiconductors GmbH, Hamburg
02/2004 - 09/2004	Werkstudent bei der Philips Semiconductors GmbH, Hamburg
03/2006 - 07/2006	Lehrbeauftragter an der HAW Hamburg, Vorlesung im Studiengang Medientechnik
11/2004 - 03/2007	Wissenschaftlicher Mitarbeiter im Institut für Nanoelektronik für das EU-Projekt *M3C*, TuTech Innovation GmbH
03/2007 - 03/2009	Wissenschaftlicher Mitarbeiter im Institut für Nanoelektronik, Technische Universität Hamburg-Harburg
06/2009 - 12/2009	Wissenschaftlicher Mitarbeiter im Institut für Nanoelektronik für das BMBF Projekt *MyoPlant*, Technische Universität Hamburg-Harburg